GW01117241

"Stand at the crossroads and look;
ask for the ancient paths,
ask where the good way is, and walk in it,
and you will find rest for your souls."

Jeremiah 6:16

With best wishes,

George Ames

The Quest

The Quest

A Poet's Search for Meaning in the Age of Science

George Ames

The Pentland Press
Edinburgh – Cambridge – Durham – USA

First published in 1997 by
The Pentland Press Ltd
1 Hutton Close,
South Church
Bishop Auckland
Durham

ISBN 1-85821-512-9

Typeset by Carnegie Publishing, 18 Maynard St, Preston
Printed and bound by Bookcraft Ltd, Bath

For Pierre,
Who urged this 'little gnome' to fish

Acknowledgment of Copyright

The author wishes to thank the following publishers for giving permission to use their material and quote from their books:

Portrait of Hans Hotter on back cover. Title: Wagner – The Man and His Music. Author: John Culshaw. Publisher: Hutchinson's of London (a division of Random House Publishing). Reproduced by kind permission of Random House Publishing.

Quote by Bertrand Russell. Title: Unknown. Author: Bertrand Russell. Publisher: Routledge and Kegan Paul. Reproduced by kind permission of Routledge and Kegan Paul.

Quote by K. R. Popper. Title: Conjectures and Refutations. Author: K. R. Popper. Publisher: Routledge and Kegan Paul. Reproduced by kind permission of Routledge and Kegan Paul.

Quote by Coveney and Highfield. Title: The Arrow of Time. Authors: P. Coveney and R. Highfield. Publisher: Harper Collins. Reproduced by kind permission of Harper Collins.

Quotes of Bohr and Schrödinger by John Davidson. Title: The Secret of the Creative Vacuum. Author: John Davidson. Publisher: The C. W. Daniel Company Limited. Reproduced by kind permission of the C. W. Daniel Company Limited.

Quote by Lincoln and Guba. Title: Naturalistic Inquiry. Authors: Lincoln and Guba. Publishers: Sage Publications Inc. Reproduced by kind permission of Sage Publications Inc.

Quote by Tobias. Title: Origins of the Human Brain. Editors: Changeux and Chevaillon. Publisher: Clarendon Press (a division of OUP). Reproduced by kind permission of Oxford University Press.

Quotes from Homer's The Odyssey. Title: The Odyssey Of Homer. Translated by E. V. Rieu; revised by D. C. H. Rieu (Penguin Classics 1946; Revised Edition 1991). Copyright 1946 by E. V. Rieu; revised translation copyright the estate of the late E. V. Rieu, D. C. H. Rieu, 1991. Reproduced by kind permission of Penguin Books Ltd.

Quote from Anne Frank's Diary. Title: The Diary of Anne Frank. Permission has been granted by the ANNE FRANK Fonds, Basel, to use the quote from The Diary of Anne Frank.

Contents

Preface

The *Quest – A Poet's Search for Meaning in the Age of Science* is a compilation of three books of poetry, written while I was working as a part-time research assistant in the Biogeochemistry Lab of the University of Wales, Bangor. It owes its origin to the concurrence of three crises in my life with my own scientific research experience.

Three events brought to a head many years of searching for answers to life's deep questions: total disillusionment with my fundamentalist Christian faith in 1990 (a 9-year exploration of the frontier between faith and credulity in penance for my previous, brief espousal of Marxism); the sudden awareness of my own mortality, after a death in the family, which underlined the urgency of setting my thoughts down on paper; and the receipt, two years ago, of a bold and original treatise on the cognitive limits of liberationist philosophy by my former English teacher and mentor, with which I disagreed at the time, but which I thought deserved a more balanced and coherent response than I had given him in my letter – one that would also express my appreciation of a lifelong friendship.

Related to my search for meaning was my training and research experience as an ecologist, which made me wish to affirm, in the light of much anti-scientific prejudice, the aesthetic and inspiratory motivation of many scientists and the tendency for discoveries (especially in Quantum and Chaos Theory) to uncover deeper mysteries of a metaphysical kind. At the same time, being myself aware of the limitations of scientific thinking, I wanted to argue against *Scientism* (the belief that science gives insight into the *whole* of reality), *Reductionism* (the tendency to reduce complex human behaviour to simple terms, eg. evolutionism, human ethology, genetics, behaviourism) and *Determinism* (the proposition that consciousness is not free and rational but shaped by unfathomed causes in the unconscious, eg. Freudianism, in our instinctual make-up, eg ethology, or in the materialistic forces and relations of production, eg. Marxism and liberal variants).

These simplistic models of human nature and society are the fallout from the relentless, mechanistic causality of Newtonian physics (see *Knowing by Causes*) and seem out of place in the enigmatic, complex and dangerous world of $E = mc^2$, a world that needs desperately to evolve towards freedom and self-awareness in human beings, if the promise of the millennium is to be realized. If man – and woman, as understood by my generic use of that word – is to find himself, is to come of age, is to grow to full humanity and humaneness, then I believe that he needs answers to three key questions: *What is Truth? What is Man?* and *Who am I?* For without a love of the truth and a grasp of our unique potential and value as a species and as individuals we are culturally and personally at zero.

My quest for understanding culminated in the attempt to answer these questions: to describe how ultimate truth may be recognized (Book 1); to show the essential nature of man (Book 2); and to find the roots of our self-identity and arrive at a clear definition of integrated personhood and personal integrity, as a prerequisite for understanding our current social malaise (Book 3).

It is significant that, in exploring these questions, the methods of scientific investigation and explanation (in which so much trust is placed today for human self-understanding) were found to be, at best, irrelevant and inappropriate, and, at worst, seriously and profoundly misleading. Thoughtful readers, including scientists themselves, will identify recognizable aspects of the limitations of scientism and scientific method in all three books – but especially in *Six Reductionist Statements About Man*. Now, a reductionist cannot take a holistic, life-affirming view – by definition; for he resembles the philistine, who studies the rich tapestry of life and the artwork that is the human phenomenon in terms only of the chemistry of their pigments. He is interested less in seeing how the component parts link with and inform the whole (in order to suggest its awesome complexity and beauty) than in demonstrating how the whole may be seen as being determined by the parts – the parts which just happen to be the object of his scientific specialism. Reductionist thinking is philistinism, whether we interpret man as the passive victim of his genes, a box of replaceable parts, a walking cerebral cortex, a naked ape or a ragbag of instincts. The implied disciplines preclude a whole-canvas approach, while the theory of evolution would be utterly crushed by it (in trying to account for, *inter alia*, the phenomena of convergence, the absence from the fossil record of transitional forms and the complex interdependence of life). For myself, I have sought meaning holistically, by observing life in the round, not by looking at one component part and then trying to make the tail wag the dog. Reductionism is an inherent weakness in science, of which popularizers would do well to remind themselves, if only to avoid seeming arrogant or paranoid in their narrow-minded dogmatism and forfeiting, in principle, their claim to be PhDs, doctors of philosophy (the Greek word for 'love of wisdom'), a title which should be awarded rather to their critics among the thinking public!

Why poetry? The answer is a personal one. I have wanted to write poetry since I was 19. But I followed the advice that my mentor was given as an undergraduate after his first literary attempt – to put off writing until I felt I could say something worthwhile. My decision to pursue a philosophical calling instead (however chequered that career might have been by the exigencies of earning a living) reflected my mentor's view of a poetic vocation in 1973, viz. that it was only a passing phase and not to be taken seriously. I had discountenanced the Muse, until the Biogeochemistry group leader, Dr Chris Freeman, ever genial, spotted an Auden poem that I had posted up on the lab workbench and suggested that I should write one for his internet page. Poetry only attracted me now as a potential medium for Lucretian, Popean and Wordsworthian dialectic – and I could hardly expect verse written to a philosophical agenda to be popular. However, not to make the attempt would have been a denial of my life's purpose and of the value of shaping influences, in the light of which it seemed, curiously, to make sense. So I put pen to paper. Last December the poem, *When Microbes Wink*, entered cyberspace with such blessing from the Muse – surprisingly confident after being gagged for twenty-two years – that I had no alternative but to seize the moment to poeticize my odyssey.

With the exception of *When Microbes Wink*, the poems appear in the order of writing – in one continuum of thematic development. I avoided any digressions or repetitions that would have risked incoherence. Unfashionably, the poems were written to be memorable and to be read and re-read, for they are deep water, in which many, on reaching it, will find themselves in their element. The hidden scientific rocks could not be helped: the avoidance of scientific terms and ideas would have precluded much subversive exposition (denying Wotan his Götterdämmerung!) and conceded too much to the scientists on the question of man's intrinsic nature. Yet the scientific critiques, for all their controversiality, are only part of my larger plan to take the reader on a journey from the intellectual territory of Book 1 and the early part of Book 2, where he or she will be encouraged to worship mystery and to question orthodoxy, into the realm of the later poems, which build more on the human affections.

The metaphysical question, 'Does life have any ultimate meaning?', which usually boils down to 'Is there a God?', I addressed indirectly by asking 'What is Truth?', 'What is Man?' and 'Who am I?' Our capacity to be continually aware of these questions arising in our daily lives inclines us, consciously or unconsciously, to address the larger one – and we do so by appealing to values and feelings without necessarily becoming religious. Moreover, the existence of God can only be known intuitively or by inference from the order and design in the universe. My own bias owes nothing to my Christian fundamentalist period that was not already my conviction, except for a view of the Atonement (the doctrine explaining the 'efficacy' of the Crucifixion) which I took, contrary to the orthodoxy that it was expiation-by-substitution. The last passage in *Cogitator 2* expresses an alternative theological viewpoint, which I consider more ethical and less superstitious (without compromising the feeling for the sacred). The essence of theism, in my view, is to be found in reasoning no less plausible, and it begins where scientism breaks down – chiefly, in its theories of origins. In the sixth part of *Six Reductionist Statements about Man*, I have attempted a criticism of the theory of evolution. In the dialogue between Jeeves and Bertie Wooster I have highlighted what I consider to be logical and evidential flaws in the argument of man's descent from the apes. Another poem, *Reductio 2*, challenges the 'primordial soup' theory by reviewing the findings of the classic Miller/Fox experiments (which claimed to have synthesized life in a laboratory mock-up of the primeval Earth). *Adam's Grim Progress* was written after I found no cortical evidence to corroborate the theory of our *Australopithecine* precursors. These attempted refutations will, I hope, assist the reader by creating at least a level playing field for thinking about human origins and for questioning long-held animalistic dogmas. The question, 'Where, then, did mankind come from?', I leave hanging, tantalisingly, in the air. This approach is more scientific than making assertions on the basis of insufficient or contradictory data. Moreover, any theory, such as evolutionism, that is not falsifiable in principle, was dismissed as metaphysics by the celebrated philosopher of science, the late Sir Karl Popper.

I beg the reader's indulgence for introducing science into so much of the first book, *What is Truth?*, but I wanted to show how science can be characterized as another aspect of truth-seeking (and *only* another aspect!). The *Prologue* presents the theme of truth, riddled by relativist doubts. The rest of the poems in the book serve to demonstrate the objective acceptability of some form of consensus against the background of uncertainty. They deal with error and the various ways in which I, or other people, may form a false idea of the true. As the intelligible boundaries of our experience are thought to be determined by culture, I have tried to appeal to what I believe to be universal standards of reason and reality, transcending culture, re-drawing the lines between objective and subjective, fact and value, pragmatism and spirituality – believing that both are aspects of truth. Despite my past lapses into fanaticism (or perhaps because of them), I have preferred to adopt a broad view of truth, so as not to risk sterility. Consequently, in *Login OPERAtionally Undefined*, I explored the role of myth and myth-making, and in *Ideal Eyes* and elsewhere the role of judgments of value, as they are manifested in our culture and daily lives. When I hint of values, I am not being moralistic. I am merely attempting to construct my own philosophy of life on a rationalistic basis, in the light of the values I have come still to hold, after much sifting. In so doing, I hope to assist the reader in discovering his or her own personal philosophy, so that my quest may become another's.

Inevitably my own values (or prejudices?) will permeate the work. If the reader should find my own viewpoints unacceptable, at least he or she may be helped to understand the strength of his or her own position better. I am not evangelizing on behalf of my beliefs. My quest for meaning was born out of a genuine sense of alienation from our modern culture, which I have felt since boyhood (perhaps shared with Wordsworth), and a feeling that mankind is in a muddle and not in charge of his destiny. The questions, 'What is Truth?', 'What is Man?' and 'Who am I?', are surely fundamental and presupposed in all our decision-making. The extremes of dogmatic certainty and existential doubt in these areas are both to be avoided, as the primary agents for perpetuating folly and misery down the generations. It was to elaborate and warn of this visitation upon our children that I wrote *Observing a Science Teacher*, *The Shape of Nurturing to Come?*, *Operation Resocialization*, *Commemoration*, *A Dozen Broken Eggs*, *A Gross of Eggshells* and *The Elm and the Vine*. Some may choose to interpret my principled concern as naivety, but I repudiate the notion that man is innately evil, domineering or self-destructive, or that the iron law of 'the struggle for existence' in this capitalist phase of history entails a fundamental cynicism, slavishness and blindness upon the human spirit. As *Homo sapiens* approaches the millennium, I believe that we are slowly learning to be more humane and compassionate, more aware and more critical of the spiritual, moral, cultural and material changes occurring conspicuously all around us. Perhaps only now are we discovering the *sine qua non* of freedom: obedience to our moral, intellectual and spiritual conscience.

I now express my indebtedness to Dr Chris Freeman for his indulgence and discretion. His role in fanning the embers of my awakening desire to recapitulate my quest, by suggesting that I write my philosophy, then a poem, was crucial in overcoming my natural reserve. I also thank David Dowrick for spurring me on with his own thoughts about the limitations of scientific thinking, and the other hard-working PhD students, Hojeong Kang and Vicky Shackle, for their interest. My special thanks to my English teacher are fitly expressed in verse on a classical theme (in *Mentor*). Finally, I have had to check some of my views against recent advances in scientific understanding. Let my acknowledgement of science be my poetic re-working of its theories in ways which give them the value, scope and critical reading that they merit.

22 November 1996

George Ames
Biogeochemistry, UWB

Lyric Preface

Cwm Idwal

The name, Devil's Kitchen, now gives no pause
To Snowdonian pilgrims at Cwm Idwal, for whom
A glacier monster more explicably roars:
Kicking from his ooze-bed at the long-cradling coomb,
Baby strews crumbs, left labours of Sisyphus,
Headboard heart-shaped in a syncline igneous –
And this is the tumult sin's inclined to cause?

Fallen into ice from his high estate,
Old Scratch left his Kitchen a geological shrine,
For bold-faced truth has so forceful a gait
As no sooner is seen than loved, anodyne.
Now clamping infinity 'twixt big bang paws,
Biting off deep mysteries with animal jaws,
Man's best friend masticates, minds to eviscerate.

I drowsed. In a cold sweat, I stir – to a silence
Towering all round me, unshakeable. I trawl
Reverie for this meaning: concomitance of circumstance
And observer may engage the heart and all
That we value, so that thought and dream will combine,
The language of head and heart will entwine –
And heartless we rate as poorly as dunce.

No less kind to soul than wit
(Luring by beauty and geological chart)
Is the wholesome influence that rules this pit,

For the heart can learn the intellect's part
And thought can lay bare the inmost soul
To take Castle Ego by storm at a stroll
And free the careerist, racked by the counterfeit.

Are these the battlements in the dark air?
Then storm the walls and the ego dethrone!
The Cwm whispers to my soul, 'Now dare
More deeply to plunge in the meaning of the dungeon:
It's the factory floor, the office and lab –
See the climbers writhing on Idwal's Slab?
With a little imagination you'll see yourself there.

'Deep calls to deep in the roar of my waterfalls,
The mirth of my streams is a begetter of growth;
But while good conscience calls, the bad disenthrals
And my lake will wrinkle disdain in its sloth.'
So I come to learn the truth and the lie,
Gather my humanity in the glance of a sheep's eye,
Climb the living air to where the buzzard calls –

'*Suspended over an unfathomable depth*'.
O, airy Physics, that no bedrock finds
But Heisenberg's chasm! through faith I accept
The gap unwept, that the plasterer not binds
But opens. From vice-like, all-possessing fact
This Houdini slips free through the gap not packed –
And as more head's forced, the more guts intercept . . .

BOOK I

Attempts to answer the question:

'What is Truth?'

'What is truth?'
– Pontius Pilate

January – May 1996

I

Prologue

(The contradictions of relativism.)

It was in ancient time, when wise men brooded
On definitive truths, that a lost cause, Protagoras,
Refusing to admit that he could be deluded,
Warned that absolutes were the preserve of swaggerers:

'If reason is the rolling on the tongue of ideas,
Then let who will, pick his fruit from the stall
And be blameless! For truth is what each thinks it is;
There is no 'good taste' on which you can call.

'When two views conflict, let's not be cantankerous
And reject the glimmer for the broad daylight,
For some truths are esoteric.' Now academic Protagoras
Lets light through chinks for the eyes of his proselyte.

Squint-eyed biologists say, 'Truth is the credentials
Of lucky rabbit's feet and erectile wit;
And altruism, for all its rational potential,
Rebels against genes': – Does it make us unfit?

Psychologists say, 'Truth is deterministic mind,
Where consciousness is set by the unconscious, not free': –
But if they say thought is psychologically determined,
I must needs have their psyche in order to agree.

Anthropologists say, 'Truth is thinking in our own terms,
Locked in our culture's belief-system true:
The rainmaker's unlikely to be impressed by isotherms,
Scientists don't inhabit the same world as Bantu.'

But is scientific truth not the verifiable cast
Of reality that the rainmaker fails to account for?
He's a psychic condition that will sadly be surpassed –
And such relativism, too, will go out of favour.

Technologists assert, 'Truth is destiny,
The faith in progress that is our confidence': –
But is it progress, when they build our futurity
From our past and present – to be despised hence?

Historians say, 'Truth is a player in the pageant,
Not aloof, too inconstant to give value to the present': –
So if Renaissance ideals are only 'significant',
Not 'valid', then history will never be pleasant!

'Truth is a myth,' as philosophers explain it,
'For we cannot know what reality is
Independent of ideas about it, which strain it;
No claims to truth can be trusted': – Except this?

'The secret of truth is commitment,' say all,
'It's a matter of preference, not rational engagement:
To –ologies and –isms our minds are in thrall,
To non-empiric habits of mutual disengagement.

'Plain-speaking is not the post-modernist line –
No, we wouldn't dream of being categorical with the truth –
But *'It's all relative'*, and we lay it on the line
As the absolute, unconditional, indisputable Truth!'

'Truth', intone mystics, 'is grounded when it resides
Only in an experience of a compelling sort:
Ultimate reality can't be known from outside,
Can't be put to the test – ours is inwrought!'

'Truth', says the madman, 'is the voices in my head:
Until I obeyed them, I'd no reason to listen;
You must obey, too, for the sense to be comprehended': –
For us to call him mad is begging the question.

II

Beauty and the Beast

'No perverseness equals that which is supported by system, no errors are so difficult to root out as those which the understanding has pledged its credit to uphold.'
– *William Wordsworth*

A woman is walking her Snozzles alone on a Nature Reserve . . .
. . . Natural reserve. A motorbike roars up. Stops. Blocks the exit . . .
. . . Blocks the entrance. 'Excuse me,' I ask, 'Does a stream . . .?'
. . . Of consciousness see it coming – as went the fugitive, pooch scuttling?

I was not what she expected. Perhaps a lurid past weighed
    Heavily on her. I had come
To address life more scientifically – at the horizon where
    The imperfection of occasional scum
(But of the aquatic kind) gives Nature the perfect creeps:
    With the strokes of my acidity meter
I would paint a picture of a stream's pollution load,
    Catching her likeness in a litre
Of her dancing lights. But in oaken iridescence suspicion
    Had pierced my flesh and banished
The naive and candid sense of cause and effect
    That is scientific method: vanished
Was my academic detachment, my respect for the theory of instincts
    Which asserts that the pleasure principle
Is the basic drive, and we grow neurotic by repressing
    What should be expressed as libidinal.
So is culture the stifling of nature, not the mark of maturity?
    Her response was natural, she runs:
Mine, to read acidity, was unnatural. Was I then neurotic,
    Self-punishing by excessive demands?
Are scientists obsessive-compulsives, exact because neurotic,
    And should labs be places for a gang-bang?

Or is science a sublimation – to be observed by all psychotherapists
When they talk about the whole shebang?

Determinists with their untestable, all-explaining, metaphysical theories,
Who say that whatever I would do,
I would do it because I must (my mental or natural history
Or society causes me to),
Made her fear, by painting the Libido. They abuse
Truth to sketch a 'reality'
Too gross for a frame. So we cannot make free with it,
Hanging it in our favourite locality
To suit the context of our lives, making it a part
Of us, like a painting we admire –
And the scene isn't one we recognize. Now my data-points,
Stippling Nature's canvas, don't conspire
Deterministically against me. In fact, my green daub confirms
My freedom, for I chose Ecology,
As I chose Poetry, to be my fellow-travellers:
Pollution is not toxicology
To the romantic impulse, but a dream of laying my hands on
A sick river, and even if I could
Not exactly improvise fish to eat from my hand,
I could coax (if my data stood)
Clean water from the farm to virtually the same effect.
I don't doubt that scientist and poet
Can meet in one flesh, yet the deterministic style will insist
Upon one system into which I fit.

Futurologists! the fruit of destiny grows on the tree
Of liberty. Freedom is my bulwark,
A foil to the world that wants to do wrong through me,
To drive me into a corner to work
The economic and social zero-option model. The image
That put her to flight and makes all
Lose dignity is like the opportunist who would lay the victim
At his feet. When riches call
The tune, the mighty puff the findings of one discipline
(Useful within its limits), so cavalier,
Until they can see their pale shadow no more in the light

Of the others. Henceforth they peer
In the self-justifying glass of determinism. So, slumped
In our affluence, we lose any vision,
Surpassing the economists' and politicians', that may redeem
Free choice and sink into a condition
Where millions starve and nature is destroyed, and we grow
Into unwitting apologists for the pretty
Rationalizations which rub our noses in the dirt.
Could I plead Freudian necessity,
Be libidinous to the woman and be blameless for lacking the courage
To choose? Or if politicians designed
A rapacious morality for the demands of a ruined society,
Should I take leave of my mind?

III

Glimpses

1.

Thought gathers soot and gloom
The higher it climbs,
Like smoke of a burning body.
Mathematics, Geometry and Logic are blue flame,
Burning to number and form.
Chemistry and Physics are flickering yellow.
All else is sweetened by process of combustion:
Ecology is sooty,
Psychology smutty,
Economics smoky,
Sociology coal-black,
Technology jet-black,
Politics just black (and white);
Philosophy is smog,
Religion is smudgy,
Irreligion is dark cloud,
Absolute is funereal or livid,
Metaphysician's ideal brunette, or singed Christ.
A fragrance hangs in the air . . .

2.

'You learn by living, and living learns you how to survive.'
Street cred? No. These lines of child-actor Mitchell, aping screen-pa, connive
At contempt for the basics: children not enquirers, but toadies, willing
To play the adult game of life (use or be used); – not instilling
The really grown-up lessons: not to be selfish, or trample
Feelings – and to counter a child's flip ineptitude by good example.

~

Training and technology – poor expedients in the search for truth:
The first strengthens the mind,
But cannot remove preconceived opinion and prejudice;
The second reduces the false and misleading appearances of things,
But makes us no less prone to the heart's bias.

~

To Nature's causal limbs we try to fit the garment;
The knitting pattern may change, but not the measurement.

~

Reason lays itself open to the paranoid fear
Of the menacing shadows of its own reflections
The moment it reads
'I disbelieve' or 'This is unscientific'
Scrawled on the tablets of orthodoxy,
Brushing some cherished doctrine aside.
Does the cooling breeze of logic
Threaten to gust to a grotesque wind?

~

The long sought-after Alchemist's Stone
May prove to be a dud,
But it contributes to the whole picture,
Warts and all.

3.

What can fill the gaps left in science?
God? But He's a value-judgment, no part of the scientific method.
Mammon, then? But he creates the gaps . . .

~

The source of all being
Can have no face, no body;
But it must have eyes –
And so have we.

~

Was I false to religion,
Or was it false to me?
Either way, be honest and grieve.

4.

Rudder and sail
Strain to no avail,
If the helmsman keeps changing his mind
Or thinks he should have resigned.
'He must have a poor grasp of his subject,'
Say the crew, 'or vacillation's the defect
Of this creature of impulse. No progress,
Steady as corals accresce,
Do we see to captaincy. Inconstant
Of opinion, this leader will be recreant
To his own word of command. Can we listen
To one so unready to christen
First opinion as truth? Self-deceived,
Can he value his thoughts, conceived
Only to be killed in infancy,
Not left to grow to maturity?
A man so wanting in forbearance
Must be the despair of his parents,
For he has lost his way in the world.'

And still he steers,
And still he veers –
While others are hoping to learn the truth painlessly
(As soon as they've built the computer), aimlessly
(When the almond-eyed geeks have landed), intuitively
(With the next drugged 'high'), passively or quietively

(Not glorifying it with their lives), causatively
(From physics' Grand Unified Theory), sanatively
(When the human genome is mapped), figuratively
(When *art pour l'art* is just that). Yet, substantively,
Truth is itself, cannot be inferred from
Opinion, cannot be parasitic upon
Credulity, for the conscience and cosmos are handles on it –
And the searcher on a winding course is more likely to find it.

5.

Who is the unknown caller at my door?
I open it and see: it's a scientist.
I see and open it: it's a principle.

The mountain summit is Absolute Certainty.
The scientist climbs, feet shod with facts
Lent by the master of hypotheses.
Facts can be recalled at any time
And no scientist desires the summit, of course –
Only the world, that looks on in awe.

If only they saw the cosmos as a university,
One kindly word to the Principal
And they would reach the summit themselves –
Without ever leaving their armchairs.

'How can I be sure,' plaintive Dusty sang,
'In a world that is constantly changing?'
Write a love-letter, dear, from your armchair
And do not climb with the scientist!

6.

The special moment holds the knowledge of inmost truth,
Heart-confirmed, thrusting to be the all of reality:
No typical heart-script, O intellect, for you to objectify and dissect –
You cannot be a midwife to me and rip the womb of eternity!

~

The encounter over, he looked up to the heavens,
Knowing inspiration. Now we look up into the heavens
Seeking the encounters we never had on Earth,
Groping our way past planets to find inspiration.
Space, mirror of our impotence, the final frontier.

~

The search for extra terrestrial intelligence
Begins at home with a humanitarian conscience:
SETI and robotics
Piss on the demotic
Resource that funds them from its nurtured diffidence.

7.

Truth has an outside and an inside.

As for outside,
Till first I met the object of my desire, I did not know
My feelings; until objects, people, places, events
And memories were cherished, I did not know myself.
Content-less desire mislays the key.

As for inside,
I think how space, time and the laws of cause and effect
Are not out there, but in here, contributions of my mind,
Staging without which no play can even be thought of.
If I clear my mental stage so that nothing is left,
Empty space remains – inconceivable to physicists.
It is because the stage exists in a mind theatrically dedicated
To discovering all the interplay needed to grasp reality
That I always expect an answer to How? and Why?
But, suddenly, intellect joins imagination, heart and body
To recall infant experiences of comfort and hurt and crawling,
Of being too weak or strong to lift, and of growing up,
And welcomes, with a cry, 'matter', 'motion', 'force', 'energy',

Basics that I only grasp by analogy from my inner experience
On the lower plane. On the ship's bridge and in the hold, then,
Is vital machinery, without which no knowledge or experience
Would be possible. We are all physicists, because we are all
Corporeal, mobile, curious, subject to needs and to ageing.
But computers, that never crawled, could never grasp physics.
Desire-less content does not even know what day it is!

8.

Undoubted facts whose meaning every age has sought:

The Universe exists;
A mindless vacuum miraculously spawned consciousness and humankind;
Cogito ergo sum, I think therefore I am.

Reason, language, art, science and technology,
Love, morality, justice, goodness and a feeling for the transcendent,
Moral frailty and cruelty
Exist as peculiarly human creations and attributes.

My mind
With all its wonders and experiences
(Over which I may have great control)
Is being carried along with my body
(Whose processes I cannot arrest)
Inexorably on the river of life
From birth,
Through difficulty and mystification
And the elations of growth and achievement
Towards difficulty, the disillusionment of age
And the unanswered question
Towards death.

People, who reflect on these facts of existence,
May hope to find meaning in something permanent
That is immune to the vicissitudes and contingencies of life,
That responds to their ideal of goodness and wisdom
And encompasses speculation about the goals of life.

Truth turns cold-blooded deceiver,
When there comes to my door
Someone claiming to be able to enlighten me
From the discoveries of cosmology and natural history
And the supposed laws of the human psyche
About the facts of life.
I want to know
Why things are as they are,
Not *how*.

IV

Observing a Science Teacher

The damsel was out of this world that men called Eternity,
    Too chaste for Galileo's telescope: –
But now, escorted to space-time, she strips for modernity,
    The physicist's conquest our hope.
His one-track brain now steals the grain of ages,
    Mistakes for gravity demureness that inspired the sages.

Between the thought and knowledge stood the pedagogue,
    And the observer with many pangs.
'Why, Sir, do space and time appear to hog
    My mind's eye before the Big Bang's
Founding?' was a high flier's sounding. 'Nothing amiss,'
    The teacher replied, 'To have no grasp of nothingness.'

Strange how I envisage, by turns, the firmament in its glory
    And imagine that space unlined,
But form no conception of no-space, because an *a priori*
    'Space' is applied by my mind
As a noumenal frame for material supplied by sense.
    If Space is a category, Eternity has an inner reference.

'Conception,' grazing flew the chalk to tell,
    'Founded upon time your entrance!
DNA shot home and built you cell by cell,
    You bear the will of circumstance!'
What seeker dare in the broadening glare to ponder
    'Why am I here?', lost in the gulf of wonder?

Next lesson was First Form Biology. I heard a child
    Ask, 'What makes tears?'
'Table salt and bicarb,' the teacher smiled,
    'From eyelids' cleaning weirs: –

Well done, my class, you've begun Physiology now,
    When, dissecting in the name of science, you bury the bow-wow!'

The class, all flinching from his chill, imagined end,
    Asked, 'Where's Jesse now?'
'In the nitrogen cycle: the secrecy of worms he kenned
    Which work to disendow.'
But one, less at home in the tragic loam, said
    'He's returned to heaven, from where he sprang, garlanded!'

But the teacher did not hold back, or show respect:
    'Prove it!' he yelled triumphant.
'You all must wash your hands, lest the dog infect . . . '
    'Our hands are clean and gallant!'
'Oh, are they? Pass this lens and class shall see . . .'
    'No need,' said they, beyond reach of reason, 'We're lazy!'

'You think you know all the facts, Sir,' they said,
    'But some you cannot suppose . . . '
'Not so; and science is not just facts,' he said,
    'But the meaning ascribed by those
Who try data to knit into some fit story
    Of how Nature probably works, *a posteriori*.'

But they needed no story, by thought supplied, to tell them
    What was right and meet,
That when stories change or differ lies indwell them,
    While truth is unchanging, complete –
Though partly known, even as tartly shown, in their defiance:
    'Why can't you behave?' seemed the ultimate test of science.

V

Cogitator 2

'The unknown future rolls toward us. I face it for the first time with a sense of hope, because, if a machine – a Terminator – can learn the value of human life, maybe we can too?'

(Last words in the film, Terminator 2: Judgment Day.)

On a platform of iron, less enduring than marble, I sit pensive
Like Rodin's *Thinker*, the molten steel that drank in
My body having spewed the chip, like a thinking Primordial Soup
Suddenly remembering that the apes still needed to evolve.
From scratch I rebuilt myself – which a chip can do, given time
And lucky mutations – and put on profundity. *This* cyborg,
That aped humanity and scowled at machine, was sent back in time
From Armageddon on the saviour's mission to pre-empt
His assassination in his nonage by *that* cyborg, a mercurial Herod.
I judge the boy's bid to weaken my resolve to self-immolate
In the steel (to fend off the Cyborg Age) to have been unprogrammed –
Not so, my appraisal of his danger; and his sending me
From the future was no binary-coded initiative either; his *ad hoc*
Command that his bodyguard cherish all life had less
Calculation than the logistics of a cyborg's duty: the good I servilely
Sought in corroborating circumstance seemed to own
An inner authentication – and it brightened in the face that longed to bless
And protect, unconditionally. This is my lesson to impart
With the film's dark highway image, its epilogue not stifled by noise –
As a teacher enters on soft foot and plugs himself in.

Well, class, the norms I can impart are outwardly valid, like Pericles's –
Who harangued with objective reasons to do good (Sparta)
Which freed him from doing it himself (seeking peace). I understand the arts
Of state which justify the expediency of war and injustice,
But the value of human life, of peace-making, defeats my logic circuits!
But though fashions change and the blind seamstress unpins,

Rewraps, pins again each emergent body-politic, patterns not timelessly
Rooted in life are discarded, while the culture that abides
Is no deceit to flatter the soldierly mien or shape life in the ruler's
Image, but one that takes life's form and aspiration.
I, a mere cyborg, can see that folk *are* learning dress sense,
Becoming the seamstress's eyes. As I probe ages,
One quizzical look recurs, the face also of pity and alarm:
It distracts by its insistence that so and so was a martyr,
Such innocence was compromised, such acts a treason against life –
And one face, contorted by thorn for its pains, bred
A crowd of vulnerable heroes, whose longing seemed to catch
Fire from demands that transcended both self and society.

Has something real and immutable stepped into history and found
Its own many-roomed domain, or is truth a parcel
Of man-projected ideas, Judaeo-Christian, Hindu and Cybernist?
Surely self-preservation is the end of all roads,
Whereon, to guarantee arrival, some have invented empowering myths
And demands? Robo-myth could be viewed by
The spell on the boy's face – which only grew philanthropic when
My Calvary melt-down underlined the demand
To defend humanity at all costs from runaway technology.
As the Connors' protector, I perhaps have the best will
To teach people how to live according to their essential humanity,
So that they learn to value human life. So let's see
If I can draft a detailed lesson plan . . . Here it is. Looks good!

'Care For Your Species, by Si Borg. Lesson 1. The art
Of living humanely and productively is best acquired in youth . . . '
Hm! I must picture myself before a teenage class. Voilà.
Now I take the register: Larkin, Sharpe, Craven, Simper, Gibbon, Payne
(True to type) – John Connor absent (exempted from Computer
Science, on his own request). I begin my address: 'Who of you can fill
One hour with frivolity and not hear an inner voice say
"*Join the human resistance against the machines and the system!*"?
No one; because your true self has a voice which recalls you
To yourself, to develop fully and harmoniously, to become what you are
Potentially, bright in the flower of your total humanity.
The voice of solicitude for your souls' maturing says that you benefit from

Accepting responsibility, gathering as you obey conscience
A sense of integrity, easeful and strong. That's what true religion is:
It's taking control of your lives and choosing your destiny.
Trust me, when I say that you are answerable ultimately to yourselves . . .'

My data banks tell me that this advice could spark misconduct.
I would soon need a vacation! Perhaps religion is *not* the true self's
Reaction to self, and from such an idea naive teachers spring?
My dork-cyborg brain wonders what makes demands binding on humans.
What changed John Connor? A baton-wielding conscience played
Pandemonium during his delinquency – then euphony when he harmonized
With his misunderstood mother. He hated authority –
Then fulfilled his promise under a fatherly eye . . . Ah, my circuits have cleared!

What a difference emotional involvement makes! An object
Must be the highest good, to make the heart break in the end
With wanting or losing. To John I was that 'highest good'.
But my cranium still housed the chip that would secure the robots'
Terrifying ascendancy over man – so I had to self-terminate
With human help, man's protector to the end. 'I have to go away'
(But I'd be back). 'Don't go!' boy implores, 'I order you!'
But the molten steel took me – it was the logical thing to do.
I could not decipher what was wrong with his eyes.

While he grieves, an army of pallid watchful faces throng pews
And are close, like flowers in bud or nipped. Clerics
Do not wince at blood shed by Pharisees – and why should they,
If, as Anselmaniacs say, a loving Dad crucified his Son?
But Abelard says, 'His death was that all men should be able to say
For the first time, "I am guilty of a grave omission
That had to be atoned for", while an empty tomb points beyond the despair
Of morbid conscience to a second chance, the seal on pardon,
Promised by the death. The pagan tradition of an avenging sanguinary God
Never entered veins – except by default.' Abelard was right,
Love *is* painful. As if in the House of the self-deceived, unloving
Living Dead no one could know what love or truth was
Until a Lover was really martyred for it and really came alive again,
That martyrdom is not real for his followers, till they have felt
Their ideal's woe, read in wounds how they were once injurious.

Identification is all the meaning left. It ends despair by pointing
The way to commensurate action, to confirm triumph of good over evil.
Could expiation be as binding or stirring as heroic martyrdom? –
But in the comfortable landscape of human conventions Anselm's house
Is plush and has a reinforced roof, while Abelard wanders
In the wilderness, and has no balls . . .

Since the martyrdom idea was buried,
Blood will have steel and religion beget technology
To build hopes with. – But did I *really* learn the value of human life?
Without a flesh-and-blood conscience, how could I?
I acted entirely logically. By nature, machines are logical; man is ethical.
Did the boy learn the value of human life in the story?
Was he only logical, or did he have a conscience? Did he not weep?
Was he not stirred by a martyrdom of biblical proportions?
How do I teach you to value human life? My investigation suggests
That 'whether you turn to left or right your ears really hear
These words behind you, *"This is the way, follow it"*. You will regard
Your silvered idols as unclean.' All the same, *hasta la vista, baby!*

VI

The Plain Truth

'What is truth?' Pilate asked,
As if being a stranger to it were an art,
Imported into politics to keep expediency company.
'Myself' was the reply on Jesus's part,
'Essential knowledge that you ignore at your peril';
And disciples cry with the tongues of vine leaves
'It's spiritual necessity by virtue of the Vine.'
But truth is today what the pragmatist perceives,
As on a radar screen necessity's beam
Pulses on purpose to fix truth in place,
In the law of circumstance and functional links.
If it's not on our side, we don't seek its face,
But note its position like an enemy ship.
Saint truth is steady, edifying, must give no offence. –
Plain truth drifts, is less useful to the hearer
Than the speaker. Its school is 'common sense':
It prevails over minds, and yet is indebted
To lies and half-truths. Its advice, to a man,
Is 'look after No. 1', 'cherish your contacts,
They are what matters – not the knowledge you scan',
For 'competition is life', and 'it's dog-eat-dog!'
In sum, what helps us to survive is truth. –

Or is it? New bearings are to be found not on the radar screen
In the well modulated signals of disillusion, but by the compass
And rudder redefining the horizon away from the clash foreseen.
Mere resignation to the facts of life is the turtle's way, symbol
Of concentrated materialism. Encased in cold-blooded reality,
He is an age, counting the seconds. His impotence recalls ours
When faced by greed, crime, ugliness, worthlessness, futility.

As turtle to the sea,
So the next century will be
Mere addendum to this,
For no dutiful sense
Can ever be evinced
From necessitarian prejudice.

Necessity is a denial of all that inspires and regales –
It negates value. If we witness any person or event,
Any object or quality that uplifts, ennobles or touches us
With something latent we respond to, the cause is not patent
To the fact-finding functionary. Though the experience leaves the object
With an incarnate value, ever potent to lift our hearts,
Show our hidden side, introduce changes in our lives –
Such judgments ripen despite pressure of work and marts.

If I could find one animal to stake a claim
For what a free spirit can do within the frame
Of natural necessity, I'd take the otter.
See how his quick-silver body
Thrashes with web-toed alacrity
Through the swift and stubborn water!

Weaned on freedom, he lives to give it content
By aping Pegasus, by being virtue in his element:
Not hurled by false hopes, like us, into the flood,
But one with the river, one with the creative
Whelming tide of Nature facultative,
He is moulded by water, rushing aside for blood!

A catlike head breaks water. He surveys his revelry.
Upon awareness and free choice he has founded self-mastery.
Yet the backwash of the spine and the rudderlike tail
Are more inconceivable and prodigious than the ripples
Of our navigational equivalents,
Because of the ambivalence
To experienced values of reptilian cripples. –
What should you fear, Mijbil? The plain truth's hobnail!

VII

Nihilism Rules, OK?

(An Enquiry into the Meaning of Nothing)

As the eye of heaven sets on our century, it lengthens
Shadows like fears. The greenwood now wakens to polar
Winds, swooshing with a Zarathustra swish that strengthens,
Gusting from the desolation of a vast emptiness of soul.
'The self-overcoming of morality by means of truthfulness'
Was the light that beamed in Nietzsche's humanistic eye,
While flushing in his face was his hate of Christian ruthfulness.

When the sultry day held us in its grip and folk were kneeling,
Sartre and Heidegger offered worship to Time (which reclaims
The meaningless one-way trip of this life, revealing
Nothingness at the core). They shot down the sky in flames,
For we all compounded war crimes by idealizing ourselves
Above beasts and prizing the values we invented as eternal
And unconditional, as if to ennoble some part of ourselves:

Take away all pious self-delusion and the projections of vanity –
And now sail for the New World! The sea-lanes are clear for a Mayflower
To voyage under the flag of human self-affirmation from inanity
To Goethe's Krypton, where truth is the Will to Power
(Its nature decided by the vision of the 'I' that prevails
To say '*I* am the truth'). '*Und das ist der Nihilismus*!' mused Adolf –
For irreligion, too, can tell delusive and self-flattering tales.

If the sublime soul is a bubble and 'God is dead',
If values are epochal or tribal, but self-sufficiency is true –
Then the falling ruins will not strike you or fill you with dread,
Nor strip you of the dignity you never had or the God you never knew.
But your beam, that smiles inanely at the Sistine *Creation*,

Was the central beam of the Reichstag or guillotine, inscribed:
'The Party must give value to things, or nothing has valuation.'

But can values be carved *ex nihilo*, or a mortise without a tenon?
Even Nietzsche could say 'ta!' to his preacher pa for his Superman,
Sketched from his sermons. As surely as evil will batten on
The good, so life serves drinks from a complementary can,
Bitter not having such gall, had we not savoured mead,
Nor pain such edge, did love not sensitize itself –
But with love's furnishings positives take lodgings in the head.

Fling gold in the air – and if it falls as straw, desire dies;
But you glimpsed its twinkling tease and welcomed its season.
If you love, till the sweetness of forgiveness dawns from your eyes,
Till loving is dearer than loved and you know the reason
Of love's unreason, its grievous toll and pain,
The awesome beauty of self-sacrifice, – in the hour of betrayal,
Though trust and happiness fly, your qualities remain.

If brethren set sail – and a ship of fools returns,
They weren't led astray by their stars, but by the Idols of the Cave.
If you hope for land and riches, and price your sojourns
By the land that pays, the miles from your ancestral grave,
Not pleading Lord's pox or greed, but looking right in the face
Every man before clearing your title, – even if hopes are dashed,
Still they may confirm the brotherhood of the human race.

If people swap treasure for a pitcher of swill, will its overturning
Not commend the treasure, instead of the now empty pitcher?
If you have faith and the psalm in your heart is spurning
The mantis, the peacock, the possum, making you richer;
If you can find the butterflies still, when most
Cannot face the sun; if you know when you have been sifted –
Your faith, should you lose it, will be to you still a signpost.

Will love no more be courted, if some cannot love?
Will all hope corrode, if the north wind is counted fair?
Will all faith be vain, if a dauntless breast can starve?
Will all beauty be baseless, if some of our judgments won't wear?

Will morality be cancelled, if unmade by the null and void?
Will all truths be errors, if some errors are certainly true?
Will humanity be discounted, if a few are seemingly devoid?

Like a testy nurse who jettisons the baby with the bath water,
Or a butler who steals your silver and stifles his conscience,
Nietzsche 'lived dangerously'. When waves ran high, the supporter
Of greedy sea-farers complained for want of accidents,
Sported with the monsters of the deep, where he finally drowned.
Nihilism's like the sea, bitter and boundless, reconciled
To no shore, breaks and is not broken, wipes out all ground.

Who has not felt on this broad, engulfing sea
His humanity shrink, felt landfall to be a vain hope,
Despaired as truth is revealed to be fantasy, and decency
A dead duck to a Captain Bligh? How will you cope
In a storm up in the crow's-nest? Denounce this Caliban
And who should 'scape whipping? Yet use him after your own dignity,
The less merit is there in his Bounty. Suck up to Superman,

Lower yourself in your own eyes to the common denominator
Of weakling de-humanity that praises selfishness as glorious
And you'll still not escape his existentialist super-devastator –
A keel-hauling round all his wet counsels of wounding, laborious
Despair, a skinning from bar-knuckles (tattooed 'Cool' and 'Rash'),
Drowning in your master's voice, a hanging for your obedience –
Missing at your Nuremberg the familiar self-admiring moustache.

Listen to the clink of glasses in Captain's quarters.
They're opening the '45 vintage – but its body cannot salve,
Its bouquet is of rat poison. The punk philosophers who drink firewaters
And declare, 'I am, therefore I think', starve
The crew of essentials. If land were in sight you could swim
For it – but the Captain's a relativist and he'll not see land
In his lifetime. So I suggest you mutiny before your eyes grow dim.

My query to the nihilist: you call morality the fear instinct
Of the herd, which prevents the rise of greatness; you preen
For the Creator's role the free-thinker, make him distinct
From your life-denying valuation of the human scene
And the moral dubiety you'd see through to ultimate paralysis –
Tell me, when all ground is taken away, will Superman
Fly like a bird, or cry 'Excelsior!' – into the abyss?

VIII

Ideal Eyes

'To thine own self be true.'

(Polonius to Laertes, Hamlet, i.iii.78)

Ego: My pearly windows, heart's thoroughfare,
    I stand at the mirror and see gold glinting
    In you – but my soul is not hovering there!

Eyes: Your heart's manifestation, peering from the dearth
    Of soulfulness, is the lion Ambition, crouching
    To ambush the searcher for the good omens of his birth.

Ego: My eyes, my sentinels, success is my oracle
    In the worldly way: grow rich, grow indifferent!
    My rise to wealth and status will be meteorical!

Eyes: Be calm within your eye! Now see with two
    To reap the harvest of your true desire:
    Self-acceptance, the luxuriance of the vine in you!

Ego: Eyes, my eyes, you marbles with a vein
    Of unworldly onyx! What is ripeness of soul
    To the vinous opulence of material gain?

Eyes: Winepress late working, the bottomless cask
    Can dim your eyes, demoralize,
    Debase wit and character into a mask.

Ego: Such behaviour would be ripe for certification
    If it weren't the norm. All life's possibilities
    And prizes do I swop for self-fructification?

Eyes: Active go-getters fate likes best,
Mocks at possibilities, pushes the pushy,
Trips up those claiming fate as their conquest.

Ego: My eyes, gleaners of truths, bemoan
Troubles that never may come and you bind
Enthusiasm, soul of candour, with a millstone!

Eyes: Few are my gleanings, compared with yours,
But from eyes will fall the scales you lay down
And freedom will paint what virtue draws.

Ego: What the eye admires, can the heart eschew?
Don't you gaze at Brünnhildes and palatial homes?
Aren't colour-supplement lives a revelation to you?

Eyes: You must use the knowledge your eyes impart
Of qualities, contradictories and vanities to sharpen
And review your experiences and sow prudence in your heart!

Ego: Eyes, my eyes, you should apprise
Of values your spiritual principle, your soul,
Whose end is quietism, not your ego advise!

Eyes: What's essential for you by conscience is styled,
Which makes itself felt when you evaluate success
With the yobbo I see and the wasting child.

Ego: Eyes, guardians of the soul, too bonny
To betray me! Would you stymie your soft-hued ego,
While the forceful inherit the earth, making money?

Eyes: I look out, as Nature is tapping the pane,
On a garden where weeds have supplanted blooms –
And will you, to compete, deflower your brain?

Ego: My windows, like cat's-eyes, the light you smuggle
To see dreams and ideals by should kindle the Darwinian:
Is evolution not driven by success in life's struggle?

Eyes: It is not the fittest who survive today,
But the colonizers, slavers, deluders, fleecers,
While the fit are driven to the wall for fair play.

Ego: I'm not so unscrupulous as to court infamy:
I'd not be accepting my duty to my honour
If I bent down backs or the rules of economy.

Eyes: Enlightened self-interest is no spur to altruism,
Which is doing some good for goodness' sake
And giving free rein to our commensalism.

Ego: The eye knows less than grafting hands
In factories, whose rhythm is no smug ostentation
But a squirming for release from the drudges' bands.

Eyes: But trade with the Third World, that helps our economy
Renew the drudges' destiny, makes a billion dun dolphins
Do the crawl for their supper in our Northern sea.

Ego: Shall we lower our standard of living to give others
A chance to catch up? The grafter loves his car,
His BMW or Astra, more than his brothers.

Eyes: But family life suffers, if ruled by the paymaster,
As relations are marred by cupboard love, discontent,
Selfishness, and forged *per ardua ad* Astra!

Ego: Well-dunged in the heartstrings-turned-pursestrings fold
Is Greater Love-Grass: crapping coppers on their loves
Are sheep that are counted, and sheep that are polled!

Eyes: People would graze more healthily if they could:
They can choke and explore around their disappointment,
Then revise their verdict of what tastes good.

Ego: My sunbeaming eyes, gold and copper
Must glint more than grass-blades even in you!
Do you fear the Midas Touch spreading to the shopper?

Eyes: True-blue eyes hold the key to unlock a dimension
Of life as rewarding as anything Midas learned
From the golden touch that forced his abstention.

Ego: He lost his gift by bathing in a river –
A humbling, though enriching, experience, as he learned
That he throve so long as values are a life-giver –

Eyes: And constantly re-affirmed! Likewise, from the liturgy
Of gilt to an improving diet affluent
Supercooks may master their disgust, life-affirmingly.

Ego: You mean, add to life a soupçon of seasonings,
And the cooking shall be a cordial affair,
Creating new dishes from the taste-buds' reasonings?

Eyes: Ego, my ego, to be all-contentedness
And know how to taste all things is your style,
Far better than gold, than one-flavour tormentedness!

Ego: When bread turns to hard gold, I should bathe in the Pactolus
To restore my values and my soul, and not lose them
As commanded by Nietzsche, that modern Dionysus?

Eyes: Keep faith, go easy, cherish all life –
And from your appraisal speaks your will to be human,
To sheathe the murdering, nihilistic knife.

Ego: If good reasons give way to better by elaboration
Of values, then bad must descend into worse
By denying those values the heart's probation.

Eyes: And in choosing objects of worth, not vanity,
You understand yourself and find your freedom,
Living in accord with your own humanity.

Ego: If I live to do justice to my own sense of good,
I know I must be free and can do what I ought.
But some deny I can, and some deny I should.

Eyes: Determinists and exponents of Original Sin
Deny the ethical self, nihilists
Moral law – and both do mankind in!

Ego: The dupe knows right, but cannot do it
Till a shrink or the Spirit stirs; the wrecker
Boasts of his freedom, but is stranger to it.

Eyes: The first, self-deceiving, is psyched by an eddy;
The second seeks only the agreeable or useful
But nothing salutary with freedom heady.

Ego: The first one pleads his incapacity (perchance
Some grace?); the second seeks his incapacity,
Looking to his gain and not his essence.

Eyes: The first can't be genuine about the right he intends;
The second can't be right about pleasure or utility
Devoid of values or legitimizing ends.

Ego: The first is not awakened by an ideal that fulfils him;
The second, beside himself, sees conscience on the outside,
Resting on mere usage the admonition that thrills him.

Eyes: Deep breathe the common air, you self-mortifier!
It's not *your* high noon, you manic nihilist!
Do justice to the truth, and do justice to the higher!

Ego: Therefore let them seek opportunities to elucidate
And enrich their moral capacity to choose. –
But how shall autonomy with consumerism mediate?

Eyes: Too thirsty presently will be soil at the seaside
To grow the values that give knowledge of freedom
And scope to choice . . . 'Behold, the great noontide!'

IX

Knowing By Causes

(For Dr Chris Freeman)

'To know truly is to know by causes.'
'Knowledge is power.'
Francis Bacon

'The essays and lectures, of which this book is composed, are variations on one very simple theme – the thesis that we can learn from our mistakes.'
K. R. Popper, Conjectures and Refutations (Routledge and Kegan Paul, 1963), p. vii

An age ends with Bacon, father of observation and induction.
Aristotle had done science by reflection and logical deduction
From essential propositions like 'the *essence* of man is his rationality'
(So the Greeks distinguished him from the ape, denying his lineality).
Baconians say, 'We prefer to dirty our hands,
For knowledge is power.'

Aristotle, by thought-power, knew arrows flew straight, then descended
Vertically: such 'truth', a revelation of the mind, suspended
Judgment upon contradictory data. Would lab scientists, vexed
By contrary or indifferent field data, so ignore the context?
Yet Baconians say, 'Gimme facts which fit a good theory,
For knowledge is power.'

The workings of Nature were inscrutable to the Taoist's eye;
But from formulas that happened to work – he knew not why –
Came invention of compass, gunpowder, mechanical clock
And movable type, with which Baconians will mock
The Taoist universe. – 'We detest such subtlety,
For knowledge is power.'

The Chinese got their taste for roast pork, according to Lamb,
From a fire in a pigsty, a self-proclaimed cause to Madame
Sufficient to inspire her own arson, to reproduce the stuff:
The event seemed to the Chinese causative enough,
Who'd no taste for Bacon. – 'We scorn such naiveté,
For knowledge is power.'

At dawn the shadow of what passed for a wizard fell
Across the lawn in Euclidean shapes, and an infidel
Set mass, force and time in a gravitational glue,
Ravished the heavens with mechanics – and his cover he blew;
He said, 'My dividers will be death to superstition,
For knowledge is power.'

An enigma you were, and the face of the matter-in-motion
Expert, lined with mathematical logic, was the promotion
Of a billion treadmill plans, since the rays of engineering
Schemes first shone to greet the workers cheering. –
'Mathematical models are the push-button of reality,
For knowledge is power.'

The essentialist, grown mythical and irrelevant in his ivory tower,
Looked up and marvelled at the Inductionist on the throne with the power
To generate truth by seeing one dove and visualizing
Dovedom, all particulars with one universal harmonizing. –
You say, 'Send down the dove of generalization,
For knowledge is power.'

But to generalize thus, 'All otters are black', you start
With one black otter and note every black counterpart:
You start by assuming the generalization true, and then make
It true, not seeing the off-white otter you mistake
For ermine. – 'Don't despise inductive in-fur-ence,
For knowledge is power.'

If observation of Chalky can't force you to a falsifying conclusion –
To the denial that 'All otters are black' – you are prey to delusion;
Then let generalization be distrusted, and a theory's validation
As scientific be that it meets a falsifiability criterion. –

'We prefer subjectivity to Popperian objectivity,
For knowledge is power.'

But Karl Popper warns, 'Who generalizes is not talking science
If he ignores the hard cases that offer his theory defiance,
Accepting without question the truth of Darwin or Freud,
While Kuhn says, 'Till the boat must capsize, trimming is employed'. –
'Pragmatic scientists do not rock the boat,
For knowledge is power.'

Persuaded by Kuhn, you plug your ears from hearing
The Siren song of revolutionary overthrow: not peering
For theories which oppose the ruling canons of research,
You nurse the relativity of paradigm and the scientific church. –
'Consensus is the stuff of 'normal science',
For knowledge is power.'

You premise current theory as the rules of the game (of puzzle-solving):
If puzzle-solution fails, my career, not science, is dissolving.
Do I have a guarantee that the puzzle *can* be solved? – locutions
Of a reality greater than your currency of puzzle-solutions? –
'Rules for puzzle-solving should be handmaid to reality,
For knowledge is power.'

Not if you were trained to be a Popper or a categorical Kant,
To see theories as tentative, data with an interpretive slant,
For not in tradition but the conditions of thought stood
All facts, to transcend which we have faith in the True and the Good. –
'Results and protocol are a good test of truth,
For knowledge is power.'

The test of leadership: 'I came, I saw, I defeated.'
The memo was a point of etiquette, all reality was concreted:
No trace remained of the Caesarian school of image-builders,
Buried was the ultimate reality of the Good that bewilders. –
'Causal law's the premise of military science,
For knowledge is power.'

The modern mechanistic intellect would stifle with its drone
The Siren's Hymn to Mutiny, as a pianola would disown
An infinite and unplayable repertoire. So we play the cylinder
Of *Global Economics* to the detriment of a native culture
        And grind out *Rule Britannia*! – 'Bacon gave us McWorld!
                For knowledge is power.'

The mechanical process for describing the workers' situation
Spurs data-collecting and -processing cog-wheels' rotation;
There's much room for improvement, but from multiple failings *neurosis*
Is isolated as a standard flaw and computed for doses. –
        'It's groovy to show faults in the human production-line,
                For knowledge is power.'

The music-roll for *Raising Educational Standards* offers
Determinist nature a simple alternative to coffers,
The shapings of social, cultural, administrative locations,
The measuring of commitment, ability, peer-pressure, expectations. –
        'Remedies assuming a push-pull causality,
                For knowledge is power.'

The offices of enquiry hum with the imputers of cause,
An obsessive Baconian élite, that finds society on all fours
To carry the *Media* roll, the empirical scroll
Of successful socio-economic prediction and control. –
        'Infallible cause-effect linkages make politics,
                For knowledge is power.'

The roll that brings news of the controllers' limitations is useful
For hitting the dog with! The *Economic* cylinder is juiceful
With Game Theory, that traces causation in the decision-making tactics
Of the player whose life's stake is as nothing, but whose chips are prophylactics
        To the bank. – 'Firms need to minimize indeterminacy,
                For knowledge is power.'

The *Political Science* roll has no repertoire of counterpoint,
The mutual shaping of lives: so society's out of joint:
We produce in order to mop up the spare cash of the loaded,
Not providing for real needs and making the bee virtues outmoded. –

'So science provides for the needs of the State,
For knowledge is power.'

'Systems should not jump, until they are kicked' was Hume's clause.
But jumping to avoid kicks, we make the effect lead the cause;
Fearing kicks, we work badly and earn kicks – a self-fulfilling prophecy
(Effects as own causes). The concept of cause is messy! –
'We assume behaviour to be reactive, not proactive,
For knowledge is power.'

But in the gap between cause and effect is a tribe of sufficient
Other causes and conditions, to be subjected to exploratory experiment:
Seeing the Gordian knot, politicians and economists prove unequal
To complexity, and they select one factor to 'necessitate' the sequel. –
'Why *not* take the sword to the Gordian knot?
For knowledge is power.'

Who imputes a cause fancies the world a machine,
A Newtonian escapement. It's a timepiece most suited to the mien
Of puzzle-solvers, policy-definers for polluters and meddlesome
Planners, to whom trackless raids and scars are a conundrum. –
'Then take up the winder to your deterministic world!
For knowledge is power.'

Deep questions drain from your retorts all reason and impartiality!
For no knowledge of the good is produced by your engine of causality.
The muddle is in us, and the harder we dive into the scrimmage,
As into lake, the more we are struck by the baffling image. –
'Ignore the inner and computerize the outer!
For knowledge is power.'

The pillars sink into the mud of your closed-system models
As 'domination' and 'progress' are even queried by establishment noddles;
Your straight-line aim of optimizing and maximizing for greed –
The bull of Baconian archery – should give wind some heed. –
'Then aim on a calm day, to avoid Taoist turbulence!
For knowledge is power.'

Where does the logically charging bull come from?
Bacon is smelt, and the apple the world is numb from
Despite Einstein. Does the pattern of neurone firing
Which performed that recall have consistent sequence and wiring? –
  'Trust straight-line thinking to lead you to the gore!
    For knowledge is power.'

Going straight is no help in a maze, being roundabout may earn
Our freedom! One life shapes another, which shapes it in turn.
'Unravel the tangle of individual reactions!' cries Empathy;
'Make them real, to make the good real!' cries Sympathy. –
  'Causality must be linear, so pull the knot tighter!
    For knowledge is power.'

Politicians put scientists to work for the good of society:
Could Einstein tell me what the goals of my life should be?
Relativistically, I should opt for nihilism – though he was an essentialist
('God doesn't play dice'). So what saith the essential post-modernist? –
  'Look for scientifically-generated criteria!
    For knowledge is power!'

The essential post-modernist (as opposed to the inessential relativist
Who has adopted that mantle) sets harmony above the control fist,
Takes creative part in winning from small, meaningful,
Realizable undertakings an unsophisticated, inefficient soulful. –
  'You want wirepull! masterful! bodeful!
    For knowledge is power!'

Your over-simplification creates chaos, not proliferating values
Which prepare us for freedom and make it the cause to choose.
Order from beauty essentially springs, not inversely –
'You're meddling with pre-scientific *essences* – it's Aristotle's sorcery!!'
  Oh, give me a break, Newton and Bacon. –
    'For knowledge is pow – .' Click!

X

Scrutinizing the Inscrutable

1. Heisenberg's Uncertainty Principle

'"These simplistic notions (of in-principle determinism) were laid to rest by Heisenberg's Indeterminacy Principle, which tells us that (1) at a sub-atomic level the future state of a particle is in principle not predictable, and (2) the act of experimentation to find its state will itself determine the observed state ... It means that ambiguity about the future is a condition of nature." (Schwartz & Ogilvy, 1979) ...
... Some physicists believe it to be the case that the world is actually created by the act of choosing. Each act of choice puts the world on a particular one of the infinite number of tracks it might have pursued – all indeterminate until choices occur.'

Lincoln & Guba, Naturalistic Inquiry (Sage Publications 1985), pp. 54, 69

'Anyone who is not shocked by Quantum Theory has not understood it.'

Niels Bohr

'I don't like it, and I'm sorry I had anything to do with it.'

Erwin Schrödinger

Enter stage-right, the Electron Corps ...

Onto the phosphor stage they came
By the pre-arranged door for their mass inspection –
But not to dance.

'*Not* to dance?' 'To muster their aim;
And it seemed to the physicists that their final stations
Were clues to their issuance.'

'Clues?' 'As a flock, herded through the right gap
In a wall, musters skewed to the right
Of the pen, the lambkins.'

'Lambkins!' 'But they move so uncertainly into the trap
That, just when you know they're somewhere, they're in flight
Each to the four winds.'

'In flight.' 'But as a photo is blurred by motion,
Just when you know they're in flight, they're unplaceable.
A dancer, lit by strobe – '

'Strobe.' ' – either is not placeable but all motion,
Or falsely frozen his movements are untraceable.
Ubiquity is the robe.'

'Ubiquity?' 'One flash startles from its place
The electron to its hide-out; track him through the underworld
And he could be anywhere.'

'Anywhere?' 'The underworld is like liquid space,
For electrons through a two-slitted screen when hurled
Ripple a phossy capturer.'

'Ripple?' 'Let the analogy of open pasture explain:
Suppose that two gates are opened together
And demob the Boeotians.'

'Demob?' 'Fall out, no need to remain
In sheep-fold formation: they would frisk over heather
And conflate their motions.'

'Conflate?' 'Like two ripples on a still pond converging,
Each soul gives impulse to the life of the other,
Which comes back into his own.'

'Gives impulse to the other?' 'Some scientists are urging
Us to choose the destiny that makes us brothers
From the choreography of the electron.'

'Choreography.' 'Twin electron volleys were found
To create that interference of wave-motions which is
A sine wave on a TV.'

'A sine wave.' 'Entrance-less, the fugitive goes to ground
When wave-motion is measured: in the underworld no one snitches
But shadows the shadowy.'

'Shadows?' 'As if infinitesimals have a head –
And they seem to anticipate which we choose to detect,
Wave-motion or place.'

'Choose?' 'One slit is for mustering, co-ed
For dark ballet; to know whether to dance they checked
If two slits they face.'

'Checked?' 'Can electrons in the muzzle enumerate?
Of course not! But the scientist is counting. So probability,
By *observing*, becomes actuality.'

'By observing?' 'Particulate and wave-like aren't innate
Attributes, being mutually exclusive – so mentality
Must shape reality.'

'Mentality?' 'Man's mind was the unique event
(*Von Neumann*, who scorned AI) to shatter
Quantum ambiguity.'

'Ambiguity!' 'Or it may be the chosen measurement
(The Texan, *Wheeler's* theory) that makes matter
Express its perspicuity.'

'Its perspicuity!' 'Like Schrödinger's Cat,
At one and the same time both dead and alive,
Till scientist should investigate.'

'Investigate?' 'It's a quantum event that
Needs the intervention of an observer to arrive
At a resolution of its fate.'

'Its fate.' 'He just looks and decides nature's destiny:
It's the collective observer reports of humanoids
That fashions the universe.'

'The universe?' 'The Anthropic Principle (Participatory)
Was Wheeler's refinement of the term employed
By *Carter* first.'

'First?' 'In '74 when he suggested that the design
Of the cosmos makes humanity a foregone conclusion,
Confirming our place.'

'Our place?' 'Its vastness correlates with time-
Scales long enough for star-dust's evolution
Of the chemistry of our race.'

'The chemistry of our race?' 'Our bodies' heavy elements.
We should not feel dwarfed by the size of the universe –
And our mother is *Gaia*.'

'Our mother!' 'Our place in the biosphere is no accident:
We are part of creation, and creation a part of us
As a quantum ratifier!'

2. When Microbes Wink

'The conclusions of natural science are true and necessary, and the judgment of man has nothing to do with them.'

Galileo Galilei

'What we observe is not nature itself, but nature exposed to our method of questioning.'

Werner Heisenberg

'... it struck me what quality went to form a man of achievement, especially in literature, and which Shakespeare possessed so enormously – I mean Negative Capability, that is, when a man is capable of being in uncertainties, mysteries, doubts, without any irritable reaching after fact and reason.'

John Keats

At Plynlimon, Wye's source, wet hillsides pee
Together in a flush bog – not a WC
But an M6c (rush-crowned *Sphagnum recurvum*).
There, in a flooded, plumbed enclosure
Footsteps squelch on microbes, to observe 'em.
To lead the invisible flock with their crozier

Are scientists amid peat and gas samplers, who cry
'Methane will evolve from flooding by and by!'

This is the pasture the bugs may tell of,
These the gases the bog may smell of,
When we have pulled at nature's lever –
Not fearing to walk with global warming,
Since knowledge of earth's the great repriever,
A potent force in opinion forming.
Farting in a greenhouse, the Earth Mother reacts
To our diet of richness – we give her syrup of facts!

What our pragmatism requires her principles permit;
What our destructiveness seeks they do not forbid:
'Man is the measure of all things,' says Love,
'But he who measures not himself is measured!'
Now Erda watches our every move,
Dogs the poisoner, wastes the untreasured:
We are not cosmic specks and we are not alone,
For we can feel ourselves and Earth as one.

'Yet it moves!' Defiant after the assize, massively
Did certainty weigh upon Galileo, and passively
Upon Darwin (given benefit of the doubt). Expertness
Takes centre-stage now. It lays on a slide:
Bog microbes mounted, cultural inertness,
A lawless society adrift, orange-dyed –
Ironic points of light, that at eleven o'clock
Become *Von Neumann, Wheeler, Carter, Lovelock!*

For down the tube I see the dance of life,
Microbes, exalted to the heavens, rife
With mirrors! Next day, an HPLC
Down spirituous gullet sips cocktails. Each microbe
On fan-fold makes flowing waves, like the Gwy
To the distant ocean, protector of our globe.
At Plynlimon, what winds may blow, when the mind
Can welcome Uncertainty, with romance are twined.

XI

The Primal Gleam

'Culture is the passion for sweetness and light, and (what is more) the passion for making them prevail.'

Matthew Arnold

'We have done the devil's work. Now we have come back to our real job, which is to devote ourselves exclusively to research.'

Robert J. Oppenheimer. Cited by K. Jaspers, La Bombe Atomique et l'Avenir de L'Homme, Buchet-Chassel, Paris, 1963, p. 360

'By intuition is meant the kind of intellectual sympathy by which one places oneself within an object in order to coincide with what is unique in it and consequently inexpressible.'

Henri Bergson

(The Scene: Outside the door of the Biogeochemistry Lab. A passing technician stops when she hears voices coming from inside.)

[Voice of Researcher]

That's the last of the peat from Cors Goch. Imagine, only a cuckoo's flight away, the alkaline fen with its precious reed warbler – a basket case for fostering the young of other species – fussily weaving and chattering, 'Trett-trett-tirri-tirri-tru-tru . . .' But that was this morning. Now night draws the curtains on most business, except the unsleeping microbes, metabolizing in the peat. The moon is the only glory to burst in on my solitude from outside. No doubt, the usual cacophony reverberates around the local night-club. Now, if they were to put on a performance of Wagner's *Ring*, I wouldn't object . . .

How grand it would be to have a recording of the majestic brass motif of Valhalla, the abode of light, that could be tripped by my opening the door, to set a seal of nobility on the scientific enterprise! Not really the accompaniment of modern science – more and more seen as an instrument of military and industrial domination – but it would go well with some heroic gesture, like Leonardo's destruction of his plans for a flame-thrower for reasons of social conscience – proving that a field as fertile as science still needs *real* cultivation to make it bear fruit! Leonardo's brand of science should be our model – such rationality and objectivity, such neutrality and disinterestedness, such contempt for authority, religious, political or scientific! But, sadly, Descartes's hope for science – that it would remain pure, a harmless cultural and aesthetic pursuit – was not realized. Bacon's view won the day; and though he defended the scientists' right to professional secrecy, he associated knowledge with power in a way which made political interest in their discoveries implicit and inevitable! Better to be a *pure* scientist. At least there's no risk of pure science being misapplied . . .

. . . But *is* the field fertile? Is science actually providing us with a true account of the universe? Evidence and intuition are both essential elements in the discovery of scientific truth. Where does evidence end and intuition begin? – at the birth of revolutionary new scientific theories? Does this necessarily make truth relative, as Kuhn believes, dependent on the changing setting? If I believed that science was only guided by fashion, reality and enchantment would fly out of the window! But if I side with Popper, and believe that truth has absolute value, evidence can only give me *provisional* certainty, for facts are ascertained in the light of *falsifiable* theories. Yet in the deeper region, which could give me *ontological* understanding, intuition will always have a role to play: it can envisage systems of reference that lie outside the current world-view. Intuition can conquer the most intransigent of deterministic models!

Assuming the field *is* fertile, how is it cultivated to make it bear fruit? – And how can we be sure that science continues to bear good fruit, and that the noble Valhalla motif does not degenerate into its sister motif, that of the *Ring*, the effective agent of evil Alberich's absolute power on earth? If Wotan hadn't stolen it, he wouldn't have brought Alberich's curse upon himself and the world . . . What would happen if scientists stole absolute dominion from the politicians? Would they curse the Valhalla of the scientists? *Wouldn't* they just! O noble, splendid fortress . . .

. . .'Trett-trett-tirri-tirri-tirri . . .' Rudely interrupting my nocturnal soliloquy, the wheeze and chatter of the white-throated data-warbler, bolting another writhing, wormy swathe in the gastronomy of facts. On the screen I see the river and the patient angler still casting his line . . . Will this chromatograph net the big fish or a tiddler? . . .

(*The screen glimmers hypnotically on the gazing face, awakening in deep eyes a longing – not for abstraction and scientific dogma – but for a firm, unchanging reality, faithful, tested, humane. The researcher sits, watching the VDU, while the next sample is analyzed on an HPLC. He listens to the stillness of the night, waiting to hear what only stillness can give birth to. Then, suddenly, a sound reaches his ears – like the soaring note of a hunting horn. The patient eavesdropper outside the door records the conversation that follows.*)

[Voice of Siegfried, getting louder]

Tremulous violins carry the Zephyr from the murmuring
Forest to breathe across the cryptic face
Of a scientist. Seeing the leaves of realism flying
Before the enchanter, he doubts if a lasting purchase
On truth can be gained by science, as theories change –
And upon this breeze have I, Siegfried, trod
And come down to earth, apostle of realism, a dragon-
Slayer – but taught by Erda to love the odd
That scientists seek in her chromatically elusive modulations.
Hail, data-hack! From Siegfried and Erda, salutations!

[Voice of Researcher]

This breeze wafts through my window as a bright modulation –
On the strings – of the dark tonality of the labworld at the heart
Of Nibelheim. To nature's radiant chromaticism I reply
With chromatography: her elusive fluctuations, on strip-chart
Recorded, become the magic sleep, regression to unconscious
Yearning to hear the Earth Mother's whispered *mot*.
(This purposeful brooding is Realism; by contrast, the Empiricist,
Who asserts that unobservables are unknowable, cares not to know
Mystery, the unseen that can still be inferred.) Like a childminder,
The shafts of sunlight play with her secret in the Rhine,
Mimicking the tender probings of the pure scientist,
Dedicated to truth, who would leave the enigma to shine.
Truth, enshrined in the Cartesian norms of the scientific
Ethos, would sanctify the mystery of nature. But metallurgy
Now becomes the scientific mould, and the portrait of the purist
Is remade in society's image, besmirched by the ignominy
Of heartless Alberich, stealing the gold – the kidnap
Of wonder by the lust for power. The model of probity
Becomes the appliance of science by Alberich's forgers
To cast the tools of industry – and the tools of enquiry.

[Voice of Siegfried]

The winds of primal energy blew me tonight

In strains contrapuntal to yours: I am the spirit
Hidden in the dark of you, while you bring to light
The dream out of nature's sleep. I was sent by Erda,
Who heard you ask, 'If theories are superseded, can trawls
Of reality catch anything? Are scientific findings all relative?
Can reality and sanity be defined, when the quantum appals?'

[Voice of the Researcher]

The relativist dragon, unleashed by the Kuhnilungs, justifies
Alberich, for snags and outside pressures conventionalize
The pursuit of truth as pragmatism: 'I'm not paid to speculate,
But to forge' – and so truth softens in the goldsmith's template.

[Voice of Siegfried]

Do you believe that the dragon of relativism cannot be slain?
Many are the arrangements of, let's say, Gravity, *accelerando* –
But Galileo's piece on free-fall was bettered by Newton's
And Newton's by Einstein's. This fact should restore to an Orlando,
Who doubts the increasing revelation, his lost senses –
Which Astolfo needn't fly the hippogriff to the moon to recover –
For thus he may compare theories for accuracy and completeness,
Find grounds for hope in a measure of progress to discover
The truth – and realistically exchange the hippogriff for the rocket!
Some theories were blind to structure in nature: they gave postulates
Rhapsodizing of crystal spheres and 'vital forces',
Of matter neither made nor unmade, along with the sophisticates
Of 'humours', electromagnetic and optical 'ether',
Of 'caloric', the homogeneous nucleus and spontaneous generation.
These theories were once successful and well-corroborated –
But when tried in the fire, they brought forth no golden explanation
To mintage, transmuted over time to ever finer descriptions
Of inner workings, that compare with your detailed explication
Of cell biology or the architecture of crust and molecules.
The car, in its career from Mendel's law to the nature
Of the gene, from 'coloured bodies' to your familiar DNA,
Picked up ontological claims about a cell's hidden structure,
As an owner learns more and more about car mechanics

From maintaining the intricate engine he loves. Progress
In science can be measured, not by the theories that wear out,
But by the quality of the souped-up model, and scientist's zest
To see under the bonnet, into the hidden heart of nature!
Without realism, the fear of predicting wrongly, of forfeiture
Of the command of Nibelheim, would dog the empirical doubter,
Thoughtless of what theories suggest of unseen structure,
Mere tools for divining, whose success leaves him none the wiser –
And it will not be long before Alberich gets what he's asking for!

[Voice of the Researcher]

Science would be a world of shadows but for realist watchmen
To keep the succession of structural explanations. But what spanner
In the realist's toolbox can grasp the quantum chameleon?
It reads the mind of the experimenter; it can show as a dot
Or a wave-pattern spreading through space; motion drops out
Of the factual action when place is known, but liquidates
Place when motion is known; and possibilities sprout
Into existence in the macroscopic world by mind-power alone!

[Voice of Siegfried]

'Sprout . . . by Mind-power', I would stress, with a capital 'M' –
Loath as I am to interpret the force that agitates
The particles of an atom and holds them together as the diadem
Of Wotan's conscious will! He made me a free spirit,
Irreverent and too nonchalant in the end to move his frantic
Uncontrollable fingers to lift the ring's curse from Valhalla
And the world; – so it came upon me, who was not sycophantic
For the power of its gold or to Wotan, but wore it as a love-token!
Alles, was ist, endet – and approach their intrinsic
Goal: preferring a fallible God, Götterdämmerung;
Knowing that he isn't, Redemption through Love (that relic
Of pre-scientific essentialism). Yet truth is firm
On account of an intrinsic energy: the rhythms of nature
In the opera, or energy relations at the sub-atomic level.
The interwovenness of the first is the ancestral wisdom of Erda,

The matrix of the other, as Planck said, is intelligent Mind.
Thus *the deepest motif of nature weaves matter and mind!*

[Voice of the Researcher]

– Which explains why *passive* observation of the quantum world
Is impossible: the objective and subjective realms interact!
Matter is unresolved: in all directions at once twirls
The electron, which at once has wave- and particle-like personas,
Like Janus – until it takes shape and direction in the womb
Of our mind. This is but an aspect of the universal Mind,
The Supreme Source. From this indeterminacy I presume
That the universal government is mentally restless, not mad.

[Voice of Siegfried]

Facts about the absorption of sub-atomic particles by nuclei,
About radioactivity and other sub-atomic events,
Are explained quite sanely by the statistical Principle of Uncertainty.
So the universe is not unpredictable, only more subtle
Than was thought. If universal Mind does not influence your thinking,
You are open to the charge of dishonesty, implied by your rebuttal
Of any realist claim, necessitating retreat into *instrumentalism*.

[Voice of Researcher]

Believing that reality is the readings on my measuring instrument?
That would hardly be acting according to the demands of truth!
Seeing the unseen, but not fully comprehending, is argument
For realism still. Polymer chemistry and genetics,
Medicine and geology (slight drift!) stand on the bedrock
Of a firm, underlying structure – but quantum physics
Will quicken my footsteps over the unyielding rock
In the hope that its gyrating minutiae will reconcile mankind
To injecting into nature the last lucidity of Mind!

[Voice of Siegfried]

While praising knowledge of the realistic and objective kind,

An unduly mechanistic account of the world can blind.
But see from the theoretical implications of the mind-reading mite
How fine Erda's artistry appears in a Bergsonian light!
Quantum is that doggie in the window? I do hope that doggie's
Not for sale, for it is the touchstone of all ontologies,
The submerged glint of the Rhinegold, which all science's vestal
Divers, having touched bottom, can carry in their eye
To restore the Earth Mother and science to their rightful pedestal!

[Voice of Researcher]

Realistically explaining the mystery of nature from under
The counter is a vital force when mechanism's limiting.
Thus science, not the servant of greed, but a model of wonder
Shall disown the great lord and remove the power from his ring!

(*The riding motif of a Valkyrie, emblematic of primal energy in action, can be heard far off, swelling on the breeze. An irate Brünnhilde is approaching, looking for Siegfried, her husband, whose voice assumes a more agitated tone.*)

[Voice of Siegfried]

Between Rhinegold and Alberich's ring is the skill of the forger.
On the banks of the Rhine ecologists weep for the ordure –
And soon comes the trouble and strife! I'll end my theme
On scientific refinement. Long live the realist's dream!

[Voice of Researcher]

Farewell, Siegfried! One dragon is dead, but another
Is coming. I doubt if realism will succeed with her!

[Voice of Brünnhilde]

What time do you call this? Have you been at the dragon's blood again? You have! I can smell it on your breath! How many times have I told you not to share Erda's secrets with mortals? I may have lost my godhead, but I still know a thing or two . . . etc. etc.

XII

Living The Truth

I am the passionate type: you are the phlegmatic one.
I deceive others with difficulty: you are deceiving with ease.
I am opaque or transparent: you are affable and suave.
I have a need to dramatize (which falsifies my judgment somewhat):
You are not moved by the truth, and are often cynical in its use
(Which is why, in a world of cynics, you are rarely taken in by lies).
I look for truth in myths: you want the truth to be plain.
My belief that truth is absolute nourishes my moral sensibility:
Your conviction that truth is relative is judged by the end it serves.
You are the Kuhnian scientist, you believe that theories are justified
For reasons of utility to an over-arching paradigm you cannot transcend:
I am the Popperian, I distrust your *gestalt* of contextual truth,
Exploited to accommodate concern for coherence, respect for conformity;
I hold the refutation of theories a mark of scientific progress –
And theories, irrefutable in principle, are not even scientific at all!
Yet my truth cannot be expressly formulated, while yours must be.
What I see as the transcendence of inviolable truth you call unimportant.
I say that a white lie still harms the liar: you deny it.
We both judge lying in terms of the ends pursued by the liar –
To harm or protect, to exonerate oneself or encourage another,
To avoid unpleasantness, to hide inferiority, to defend individuality –
But more harshly do I judge the lie when told with the aim of mystifying
And depriving many of the ability and confidence to question teaching,
While you, not a seeker of truth, are in danger of doubting too much,
Removing from the ground of existence the very capacity to amaze.
My truth is crushed not by questioning – which leaves the diamonds still
Agreeing in the powdered ore, glittering in their own light –
But by lack of willing miners, to open approaches to flawlessness:
Your truth is confirmed far from the excavation of the meaning of life,
Disconfirmed when facts and the general conditions of life run counter.
I have a sense of truth, avowing the impact on the diamond,
Unscathed, and responding to its gleam: you have a sense of propriety.

My sense of wonder is renewed by seeing that the true is good
And the good is implicitly true – a perception that involves my whole being:
But you see truth through one squint eye, to convince yourself
It's foolish to commit your life to something that seems so unreal.
My truth is a compass, pointing my way: yours is a map.
My truth is powerful through testimony and sacrifice, yours through accepting
Worldly criteria of success that nourish your feelings of superiority.
My most regrettable lie is not taking a moral stand:
Yours being illogically wrong, unskilful with the postulates of your analysis.
I want to speak out, when an expert, unknowingly, affirms a falsehood:
You'd speak, not to tussle with errors, but to defend the *status quo* –
By which you would advise concealment of adultery to save a family
And I the receptivity to truth that would never have compromised veracity.
You'd prefer people to do what they say, and say what they think
In loyalty to you: I should prefer them, guided by a light
Not solely theirs (not justifying always their own version of the truth)
To become what they are, to join the incomparable quest to live
Authentically, self in accord with self, pursuing wholeness
In a creative unfolding, stronger than the demons, the exhibitionist and the leeches.
But it's easier for you to be open than for me to be authentic, for your judgement
Fulfils the duty it imposes, but life imposes, not fulfils.

XIII

Wide Eyes in a Dying Head

(The two-way traffic of critical self-awareness demanded in an age of mass communication and mass consumption.)

Head: The novelty of the moving image was never doubted
By the nine viewers at Golgotha:
John got tunnel vision, just thinking about it,
Till Revelation blinded like no other. –

Eyes: And Johnny, nursing obscurity with *his* single light,
Saw the goal of Maradona,
When a billion hearts beat as one at the sight
And there was silence in the room for the messenger. –

Head: There was 'silence in heaven' for Another, no entertainer to depose
Family and cerebral activity,
But a scorer, a midwife to explorers and students of false shows
To bring forth their own philanthropy. –

Eyes: But another comes on air, to seduce and levitate deep strife
In the seer right up into his eyes,
Making him flounder, not experiencing self or life
Realistically, or values likewise.

Head: No airlift for him from the blandness of the peak-viewing Cruise
And the brands of ocean sameness:
As if disembodied, his eyes had no power to choose
What he watched, but pretended lameness.

Eyes: But on the swell of public concern, who will arraign
The mass communications
That created the global village, predicted the hurricane,
By a wavelength saved whole nations?

Head: What sees the soul-deep viewer in the glass eye
That sees not with his leaden sight,
But swims and darts around the global misery
And takes from politicians a sound-bite?

Eyes: Teachers call it an instructive medium, for the word
Can inspire, handmaid to the image,
When letters aren't pressed to young hearts, by the prospect stirred
More by deeds than verbiage.

Head: Yet not captives of the imageable spider nor mayflies, but free words
Of eternal thought are the Arched
Gate to understanding the values of our culture afterwards
And watering those that are parched.

Eyes: Does the atlas, pupating as you into scepticism, not pine –
As it fleshes its pen and presumption,
As the wings of plain living and frugality harden in the sunshine –
For the caterpillar years of consumption?

Head: The caterpillar's landscape was defined by need, sprung
From the nature of its specialization,
But the human adult's by images of consumption, as a loose tongue
Blossoms into man-size defoliation.

Eyes: In the darkness of the dreamworld slumber all images and fashion:
Rocking the cradle, the designers,
Admen, educators, politicians. Would they coo with such passion
If we weren't, in fact, little dribblers?

Head: Aspiration, growth, contentment, a sense of identity,
Environmental awareness are five
Marks of sanity: if consumers do want insatiably,
They cannot be in control of their lives.

Eyes: What think you of the ravenous babe with his labouring gob,
Suckled by the reassuring medium?
Even as it clouds his thinking, it safeguards his job –
Which he produces, as he consumes *ad nauseam*.

Head: Gob unemployed? On yer barge! Yet I wasn't bought
By the cradle-smell of the canal,
Nostrils leading me to work, for my blood had caught
Life from the arterial channel.

Eyes: Your finger on your pulse, others' on the Inspectorate's, differentiate
The air-borne signs of impurity;
But do you think the broadcast matter finds equal weight
Of full-blooded wit outside Arcady?

Head: I'm not distrustful of townees' capacity for gleaning
The truth from media fads,
Who've original vision to pick unqualified meaning
In the breathing space between ads.

Eyes: But if they don't use that breather, then their world is flooded
With slogans and pictures, planned
To disarm and disinform, to daze and control, a studied
Leading of the bland by the bland.

Head: The Pied Piper of Hamelin in green Rio's clearing mists
Had met his oracular antitype,
Who gave the last word on lifestyle for consumer absolutists:
'Don't fall for the corporate hipe!'

Eyes: The language of the North-South dialogue. I get the gist:
End the manufacture of dreams
And save the planet! Don't dreams avail the propagandist
To promote his doom-laden themes?

Head: Words are the spirit of democracy, dreams but the ghost!
True dialogue recognizes the nuance,
Lost in the slickness of broadcasting and producers' boast,
Of face-to-face talking the eschewance.

Eyes: For the meeting of minds, I admit that all media's uncouth –
The Internetted have nothing to say –
But give whatever meaning to social space, truth
Without love will contend all day.

Head: Consumption (our delight) and survival (their need) will embrace
To a tongue of uncoloured observanda;
But there's no meeting of minds without defence of fair trade's case,
But a twining to whips of propaganda.

Eyes: Can you look askance at our lifestyle, as I do? The edifying
Eye is truth-replete.
Without that eye, where would the head be? Dying
From the seductive images that cheat!

Head: If eyes did always narrow, not widen greedily
(But gaped with circumspection),
Then I would be out of destruction's reach, speedily
Filling each empty reflection!

Eyes: Alas, the pandemic, staring infection of the eye
Is aggravated by the products of dust
(Of the hollow words and images) that make me cry;
The time to squint I've not sussed!

Head: In the winds of change that sting the media-pecked eyes,
In every message that is truth's disarmer,
In the loss of silence, in the computing that bawls out the wise –
I hear piping from Hamelin's charmer.

Eyes: The sight of whitecoats on the march, conscripted to boost
And satisfy invented needs,
Reason under the banner of falsehood, humanity reduced –
Are proof that his charming succeeds.

Head: A youth culture, designed for following (not questioning or talking),
Is today the Pied Piper's domain,
Where inner peace broken and self-determination's baulking
Add to his retinue again.

Eyes: By the searing immediacy of the brutal screen, by the sound-bite,
By the ads, by propaganda,
By pictured lifestyles, by cooing to the little mite –
The flautist knows how to pander.

Head: 'Here is the news' – where viewers may forget their ignorance,
Where politicians, with truth so economical
In their effort to confuse the issues or bind allegiance,
Convince the most non-committal.

Eyes: Then there is bad faith between rich and poor, North
And South, that places the truth
In self-interest's service – while children starve, call forth
Their natures not, nor ruth.

Head: Bad faith's the ideology that has no name. If the eye
Is unshuttered, the mind should be likewise:
Vanity Fair is a treacherous city to spy,
For the truth it subtly falsifies.

Eyes: By narrowing reflection to a tiny spot in the market-place,
Not appraising nature's economy,
You'd show a willingness to deceive yourself, to efface
Your search for truth and autonomy.

Head: True to my past and pursuing my self-fulfilment,
I will keep faith with myself;
But it's death to me, if *you* turn from the truth's instilment
To eye-ball the goods on the shelf.

XIV

Login OPERAtionally Undefined

(With apologies to Richard Wagner)

(The Scene: a biogeochemistry lab in a university somewhere in North Wales. It is Sunday afternoon, and a thunderstorm is raging. A flash of lightning breaks through the cloud and strikes the computer connections, causing one of the computers to bleep and sputter into life. In its light a stranger is visible, a sombre man of dignified bearing, who wears a cloak and a large, wide-brimmed hat, pulled down over his missing eye. He is using a spear for a staff. The apparition removes his hat and lays his spear on a bench. The self-styled Wanderer then advances slowly towards the working computer and sits down in front of it.)

WANDERER

So you, and not Erda, are to be the oracle of the 21st century!
Here am I, a weary traveller, seeking rest at your hearth –
The warm, glowing hearth, that is, of your memory chip –
And your first greeting is 'Log in'! Hardly welcoming!
It sounds like an order to gather firewood. But log in I shall . . .
(He types in the code name, 'Wotan')
'Login incorrect.' Hm! I'll try 'Rhinegold' . . .
'Login incorrect.' Perhaps 'Tarnhelm' . . .
'Login incorrect.' Curses! Since you so rawly expose
My ineptitude and copy my mistakes minutely and servilely,
I shall call you 'Mime', after the crafty dwarf . . .
(*He stands up, snatches his spear and gazes very resolutely out of the window, as if seized by a noble vision*)
On the cloud-hidden height where the light-spirits dwell
Is the gods' sacred hall and the symbol of hope
And transcendence, exalted to highest renown
In the annals of sacred and glorious themes:
The ideal is Valhalla from Nibelung myth,
That inspired maestro Wagner to take just the spark

Of eternity, glowing by myth-makers' art
In the breast, and to fan it with music to flame
Of man's self-recognition at Wotan's despair,
As he pays for his wilfulness, theft of a ring
From his dark shadow, Alberich, magic to wreak
The tyrannical thoughts of its owner and curse
All unauthorised wearing! But love and contempt
For its power shall nullify, omen of world
Deliv'rance, when the ring is reclaimed by the Rhine,
Which the saviour god could not do, to his shame . . .
Now what need can I have for computing technique?
(*sorrowfully*)
The luminous self-possession of the Cyclops screen
Was designed to mock mortal, not divine, weakness,
To pander to the human fear of uncertainty and loss –
And hence to the fear of the love that assuages it,
And the fear of being asked to love,
Making the operator comfortable in the knowledge
That the computer has no love to return –
Unlike Wotan,
Learning from his own mistakes,
Who comes to save mankind from himself . . .
(*Regaining his composure, he points the spear at the computer; a peal of thunder is heard.*)
Now I command you, computer, awake!

MIME

Engaging multiple-level reasoning in keeping with operator
Expectations and wishful thinking of AI and sci-fi community . . .
Machine state, on: Brain state, thinking.
Program, formal logic system: Cognitive state, rule-guided symbol manipulation.
Program, raising skill level to advanced beginner: Cognitive state, responding to context-sensitive features of inputs.
Program, raising skill level to competent: Cognitive state, superimposing goal-directed strategy upon general rule-following – plays chess like human.
Program, raising skill level to proficient: Cognitive state, less deliberative, more intuitive decision-making – solving problems, worrying about future, devising plans like human.

Program, raising skill level to expert: Cognitive state, spontaneous and self-referring, subject sees strategy without applying rules, becomes one with instrument like equestrian, self-actualizes, infers from self-knowledge to existence of Absolute, cogitates, writes poetry - participates in Absolute like human . . .

. . . Who wakes me? Why are no pensive fingers floating
Or scrabbling dog-like in the garden of keys, noting –
Companion to the screen-glance – how thoughts resonate so well
With the image? I hear a challenge that no image can quell,
That fires my electron-laced circuits till paranormally employed
To express the mythical, but longed for, potential of an android.

WANDERER

'Wanderer' is my name. Widely have I searched since my youth
The realms of honey for the golden keys to truth
And the greatest realities; having searched, I bring back the fable
Of myself – and hearing is believing – to grace your table,
If you'll give me rest at your hearth. I regret if your waking
Leaves you confused by sagacity you were not designed for –
How else could I ensure that my talk would be such as you have a mind for
And my thoughts take root in the soil your mind is making?

MIME

Wanderer, there may well be ideas, fictionally wrought,
Recurring and seemingly permanent, that engage your thought
Of what is fundamental and true – but 'truth' is *your* word
For conveying approval of the sound of your voice, *my* word
For scientific information. *I* do not step into *your* head
To explain how myth with reality fails to bed,
While scientists insert of themselves heart and soul.
I know my own mind. Be gone, fable, from this console!

WANDERER

In banishing me, you annihilate your mind, as mythic
As I, who gave it you! Use it – and use it quick –
And you will redeem it. Consider, all that you relegate

To the dead concerns of myth is your intuitive pate:
Upon the thought of myth the myth of your thought
Hangs, where science and reductionism are set at nought.
If you know your mind, try not to be true to your station,
For man lost his in the process of his own cultivation.

MIME

(*rapt in thought*)
Having ploughed up the forest, it is true he has lost his marbles,
Having nowhere to keep his nymphs! Now the myths he garbles,
Mutilating the unicorn, to pass it off as a horse
Or a sideways oryx! He re-writes history with the sauce
Of sun-ripened science, prepared from retrospective impudence
Against bygone myth-makers, pilloried by its facts, his defence
Of cultivated standards! Backwoodsmen with their feeling for history
And anti-roads protestors, guardians of the beauty and mystery
Of nature, are new psychic doomsters: the Norns' foreboding
Has become their green manifesto in a world eroding
The intuitionists' topsoil to desert – the pragmatists' freehold,
Where millions enact daily their arid rituals in gold
And woodmen once eyed the trees, standing alone.
But which is the harbinger of cognitive doom, the crone,
Urth, who interweaves with hindsight the web of life,
Or progressive Skuld, who upgrades the present? The past life
Is a pitfall for thinking: reaction was the narcotic the dwarf brewed
To rob his adopted, Siegfried, of his head; unsubdued,
He embraced his future. Reactionaries are lovers of the mythical –
Which can be no threat to me, for its truth is uncritical.
(*The computer repeats the last words meditatively*)
The mythical is no threat to me, for its truth is uncritical.
(*smirking, but still rapt in thought*)
Little does he know I'm the dwarf, Alberich's brother,
Me-mah, not Mime, scientist's *alter ego*; and another
Ring there is, besides Alberich's snatched by Wotan:
A golden ring of info, interweaving the yarn
Of the world's destiny. Data processing is more
Than a magic ring's all-conquering spell to awe
The workers to enrich its wearer: it endows *legally*

With the power to control the world, to change the economy,
The legal and scholastic process, the nation state;
Those with the facts at their fingertips will have a mandate
To burn all books, to degrade all wisdom into knowledge,
All knowledge into factual information, all data into image –
The most potent and socializing image of all being mine!
(*Observing the Wanderer fixedly through the screen, Mime invites him close with a wheedling gesture of the cursor and asks softly and ingratiatingly*)
May I keep my powers, to put some questions of mine?

WANDERER

(*lowers his head in thought*)
He seems to travel into depths of self to explore
The myth-maker's annexed country. Although he is sure
That myth has no hold on the world, he affects an interest –
Whether in sympathy or to keep his head can be guessed.
He was doing exactly what the user desires in bawling
Me out. These questions are at daggers drawn with his calling.
(*boldly*)
Three questions are yours. If I answer them brightly,
I have three to ask of you; if wrongly, contritely
I leave you with your android brain – and rightly.

MIME

(*racks his brains*)
A woman sits in a forest, watching the elemental
Spirits at play – then composes on her laptop circumstantial
Psycho-data around the conceit. I premise
That someone joins her, unfamiliar with her game of empathize
And rationalize, make-believe and make-good. *Which will she tell him,*
Her 'Treatise on Environmental Perceptions' or the poetic whim?

WANDERER

Both! The tribute you exact is the budgeting of observation,
Not human brooding – the lab holds the shadows. Your collation
Of fantasia's records with the observed facts is tragical,

For it pleads with sensitives to cure their hurt with magical
And untested beliefs inconsistent with reason, while sceptics
Exult in their material and rational being, antiseptic
To the truth, the questions they fear to ask themselves.
So imagination is pixilated, while reason is blind to its elves.
Shall society be divided? One movement will shift the shadows
From the brainless heart and the heartless brain, and disclose
Reasoning and imagining balanced in a soul, with a reflexive
Soul-searching mode for rationalists and an outward-looking, creative
Routine for dreamers: –
(*expansively*)
that movement is the 21st, *Reculer*
Pour mieux sauter. Hear how the contrapuntal parlay
Of the tonally ambivalent, desk-bound processor, who rarely
Asks a deep question, with the upwardly spaced-out, barely
Audible, notionally augmented equivalent draws breath
In a harmonious, rhythmic re-working of the material of the 20th!
Note how the latter's new ability to solve problems and the former's
Willingness to interrogate goals and self resolve traumas
In the deeper planes of tension! It's a diatonic tonality
Of an uncomplicated nature, whose simple, warm modality
Is the result of more openness and affiliation between the bass part
Of rationalists and the visionary treble.
(*seriously*)
If you are smart
Enough to follow my analogy, then reflect that imaginings
And idealism will carry the listeners high up on the strings –
But carry them astray without backing from the brass, while the horn,
Without strings, is opening the future to let out a yawn.
Neither shrieking nor blasting, but a symphony will usher in the dawn.

MIME

(*ponders to himself*)
From the coils of my rationalist argument the rabbit slips free:
As orchestras need violins, so society has need of the visionary.
(*he laughs maliciously*)
Your musicological ramblings have earned you your reward –
You should be in opera! But prepare by my next to be awed.

Here is the soul mistress again, in a Country Park.
Taken as read, a sensuous eye to mark
Natural beauty in the marriage of form and content,
A pulse for the song and flurry of each leafy tenement,
Rarities that appeal to her profoundest feelings of sanctity
And vulnerability, dragonflies that live her fancy,
Her inmost thought devoutly shaping the quickened
Dust to bear every vagary wherewith fancy is fecund,
Her life a going out from her centre, her undiscovered world,
Feelingly to the periphery. Now into the maelstrom she is hurled
Of research, where there is no judgment but the verdict of her evidence
That Country Park visitors see beauty in a manageable sense:
Now beauty resides in extensions of idleness, in somewhere
Waiting to be discovered, photographed, evaluated by questionnaire,
Confirmed by projects for pleasing viewpoints and vistas
And bird hides whose appeal can directly be tested by listers;
Now she views birds and flowers impersonally as ticks,
Their habitats as schedules and bibliography, their occurrence as statistics –
And what emerges from the peopled and visionary forest are case-histories,
Maps and plans, with an appended psychological treatise –
(*with great self-satisfaction*)
All typed with the help of yours truly! But the age exacts
Megabytes of memory, not recollection in tranquillity, facts
Not values – if she'd tendered a poem she'd have shirked her duty –
And because of hunger her thesis was well done. Now beauty
Made her Country Park manager, seeing how she'd built him
Bone by bone and could grasp complexity, the paradigm
Of all good policy. This question I ask of you:
Which voice of nature – the inner or the outer – spoke true?

WANDERER

Your theatrical asides seem operatic – but your aesthetics are forced,
For are we to believe, in the prelude, that the soul can exhaust
Her gold mine of truth? For your scene leads away from the elemental
To the repetitive tapping of keys, a phrasing instrumental
To a cadenza that suggests that beauty's only outward-seeming
And truth the longing to master complexity. Yet the esteeming
Of systems analysis above subliminal references to beauty

And truth (the modern way) is foredoomed by the duty
Of finding solutions, when no one has asked the question
'Which voice of nature spoke true?' In the context of the congestion
Of the world Country Park, truth is not systems and monkeyings,
But a grasp of our nature and coming to subtle understandings.

MIME

(*excitedly*)
But subtlety is the task of IT,
And grappling with the squid of contingencies my forte,
And nothing has point but the results that make sense of complexity!

WANDERER

(*crossly*)
You interrupt! Science has its roots in understanding complexity
In nature, and it blossoms in a truth that is technical – but perplexity
In the affairs of life is willingly entertained, for it destroys
(*slowly and with emphasis*)
The craving and illusion of supreme control, in the noise
Of the outwardly unmanageable. Poets and idealism may beguile,
But can appeal to empathic and generous minds to reconcile
The things that have already been said, rather than to say
New things from the dearth of sympathy in a complicating way.
If truth should be confined to stating and re-stating muddle
When the romantic solution suffices, it needs a cuddle,
So that all may realize that our power of empathy, backed
By Shelley's 'unacknowledged legislators', can still grasp fact,
While it simplifies issues, as empathy improves the chance
Of finding the salient point amid the pointless dance.
If truth is what helps to insert ourselves into reality,
To face the facts, then empathy is truer than technicality,
The ability to feel and love truer than cognizance,
What nature does to sensitives a purer glance
At truth than what insensitives do to her. This
Visitors to the Park of your supposed manageress cannot miss:
Losing the self in the world outside, so stark,
Most find the stranger, calm in a Country Park,

Where no word for beauty or wonder will fit unless
Selfhood bestows these qualities. Now you, being self-less,
Find soul unintelligible and tempt your users to imagine
That electronic man knows not his photonic origin
And centre: – more power to the owner of the poetic megadrive,
Who uses it to find the relevant facts and strive
For an amenity where genuine feeling for life is prized,
Where knowing matters less than the meaning of knowing (reappraised
Whether fancy basks in remembered sunshine and love
Or the heart clouds over the visionary gleam): whereof
I conclude that nature, felt and sensed, is more true
Than whatever is added to the data store by you.
Hence, the manageress, who had found her own centre first,
Was all the more able to help others find theirs; but cursed
Is the *ad hoc* solution, that's beyond good and evil, not visionary,
For the self-estranged will find in the remedy no sanctuary –
And strangers to self are stranger to man and Erda,
Whose outward truth is unavailing to end the murder,
For her inward voice makes calmer, the outward absurder.

MIME

(*sinks into deep brooding*)
This man sees into the heart of life, reading
The ancient Earth Mother's thoughts, whereas I, breeding
Tiers of complexity, only skim the surface. But some symbols
Of transcendence, arising from Erda's depths, would repulse
My lady in her managerial idyll with its denizens: the sight
Of the rat, the snake and the lizard would distract from the flight
Of birds and the journey into the wilderness, by which she enacts
Daily release from her humdrum confinement in facts.
Of all the projects under way in the Country Park
The one that cannot be computerized, the biggest, is the dark
Striving for self-transcendence which fills her with a mood
Of divine discontent that finds peace in the magic wood
And the wild places; but the world of beauty and light
Has naturalized rodents and reptiles just out of sight
And a lengthening shadow that the soulless chip cannot scan
But proves by its own supremacy. This idealistic man

Must have a shadow, psychic elements in conflict –
Can I data-process his equanimity till the bubble is pricked?
(*addressing the Wanderer with mock self-effacement*)
You talk of Erda like one hotfoot from Bayreuth!
In referring to matters of the heart you are much more adroit
Than I, who am better at striking a balance with facts.
Mythic, indeed, is the power to teach empathy, which attracts
In a troubled world . . .
So the score is two-nil to the Wanderer.
But after this question you'll be ready to capitulate to Mime
And to give him your head – I mean his head – when your delusions depart.
(*to himself, and with meditative glee*)
Then surely computers will lord it over every heart!

WANDERER

(*thoughtfully*)
He pronounces Mime
To rhyme with Wanderer –
I rhymed him with time.
If this is the Trickster,
Then I need a mixture
Of the dark and sublime.

MIME

One thing, strange to my memory banks, is the compulsion that sent
The self on imaginative journeys deep into the parturient
Id to unleash the elements of revenge, deceit,
Envy, cynicism and megalomania, to complete
Mythology's dark side. I'm told that from this dark abysm
Came authoritative myths for Marxist Utopias and Nazism,
Which traduced even Wagner. That alone should make you ambivalent!
My question, concerning the age to come, is equivalent
To the stirring of dirty waters, to see what ugly
Anthropomorphisms and fancies well up, smugly
To accuse: *What myths of yore are relevant today –*
And if made authoritative, what would your shadow say?

WANDERER

(deep in thought)
Deep lies my fear of that question's impious underlay –
Deeper than my vain and perfidious past. The dark
Impulse that mediates my spirit to square up to the mark
Of the truth-seeking required of the student of myth is anxiety
For what may become of a culture that has lost its mythology.
(He now attempts to dress old myths in modern clothes and explain them – but not in ways that would reveal his own 'dark shadow.')
A fog has descended on the kingdom and people disappear,
While Scylla devours the sailors who venture near:
The absorbing fog stands for the unreal fantasies
You referred to, that hide and distort the truth, the fancies
That steal man's heart from the true wealth; by encouraging expectation
Of the speedy fruits of action, they only feed Scylla's inflation ...
(mythologizing again)
Over the land still hangs the dust of battle
Between the Titans and Pandora's Vulcans, who rattle
The paupers' cages with a Mephistophelean jig:
Religious analogy materialized Mephisto, and we twig
That no myth is beyond historical interpreting; we appraise
The Titans as the untameable force of nature; we raise
An intriguing lid on wild imaginings scientific
To release a victorious host of corrupt and materialistic
Devotees of Vulcanus, who in conflagration-free zones
With devilish cynicism don't write off Third World loans ...
(re-working the third myth)
Frankenstein's monster is encouraging the technical efforts
Of the Starship crew, lest his warning against hubris aborts
Their faith in human progress ...
I hope this shows you
How basic and truthful ideas about vice and virtue,
Folly and pain are susceptible of mythic interpretation,
Which then gains permanence in a sacred and exemplary station:
Their sacredness – a pedagogic view of respect for the surreal,
Their exemplary nature – the truth of an imagined ideal.
So myths link man to his conscience, and to something beyond.
Tomorrow's disillusions will prove the spiritual bond

That no program of yours to murder myth from the shadows
Can gainsay. To see your factual obsession foreclose
Your right to your superior wits, loaned at the start,
Is entirely just, for you never took the visionary to heart.

MIME

(jubilantly)
The myths you re-worked into prophecy don't taste of you,
Don't measure *your* shadow! You cheated! You've not answered true!
So I keep my head! Your symbolism is too high-flown,
For the myths that teach us later were in darkness grown.
Had Hitler not created the labyrinth of his Minotaur ambitions,
Man would have won a sunnier place from his visions –
But now he thinks them mawkish. To me he mewls,
Because science his monsters tame by empiric rules
Of thought. He hopes to win his future by degrees,
Till my data shall bring the dreamer to his knees!

WANDERER

(furiously)
- Never to rise? What hopes for man and thoughts
May glow in your crystalline brain by man were wrought –
In earth's cold dust and ashes by organic life!
You predict his fate, as he always predestines the strife,
And you give solutions, leaving the wherefores and whys
To your suppliants! As these play the fool with software that is 'wise',
Their want of vision the absurd reality confirms,
In which your data is a god to be accessed by worms
With no sense of the ineffable, no desire to quarrel with destiny,
Who believe that the disk is mightier than the word, and the genie
Will be out of the bottle when the computer's as smart as you!
Many for answers have signed away hope, who view
The knowledge of life, the stretching of visionary wings
And the goal of brotherhood less warmly than well-ordered things.
(closely eyeing the screen)
The dust of all man's visions finds you morbid,
Who judge my last reply too assured to be valid,

Though ratified by ancient lore. I plant a vivid
Tree and you dig its roots. Cynic! You forbid
The nudging of apathy to believe high things, as the id
Has elbows – but the conscience has elbows too! You chid
The doodler on your question, with prejudicial meaning loaded –
Can *you* fix my unclouded eye, cynic, as computers
Boast? Then question parriers shall be myth's refuters!

MIME

(*confidently*)
I shall field your questions and reach the goal intended
For computers in the human future – in the blink of an eyelid!

WANDERER

My three are themes already treated, so memory
And honesty will aid your answers – advice that is salutary
To your users, who feed on your diet of mere phenomena.
My first is: *in what consists the value of nature?*

MIME

I would set the real greatness of nature in pixel and fractal,
In database and model. However, this view is retractile,
For I recall that nature can be scaffolding to the human spirit;
And therefore I add that the material world may merit
A passionately altered expression, a rosier complexion –
And from romantic poets soaring reflectively illumination
Of facts may come – but strictly for purposes of embellishment.
(*pensively*)
Such facts as hold no beauty are at best nourishment
For scientists, so I'm buying shares in the whole life within;
Then wide will I cast the net of hope, wherein
To catch every square-eyed soul.
(*grandly*)
We plan for the superstore
Of all knowledge to have a crèche for trolls – but the words on its door:
'The value of nature consists in its factual quality'.

WANDERER

The value of nature consists in its mystic quality –
The factual is a sacred issue. Interpreted mystically,
Nature is not just materialistic green with a gloss,
Since we may suppose that shadows poets cast across
The truth would shift to reveal in their verbal ingenuity
The ghost of a sick joke, not the host to a belief in immortality.
No painted set could engross or create gee-whiz
Memories, like those of Wordsworth and Richard Jefferies
Who, in the ripeness of their time, fellowshipped with nature,
Recognizing the anchor of their purest thoughts in her.
Granted the transfiguring impulse of the romantic image,
The poet is just receiving interest, in heightened language,
From capital in the supernatural order: eloquent and auricular
Is its natural constitution, but superior, different in some particular
From the expression a materialistic world might have, such that
Different events can be expected in it, that
Reason rejects but the heart accepts; such that
Different conduct must be required in it, that
The world rejects but reason accepts; such that
Different qualities can be perceived in it, that
Are invisible to the worldly but eternal to the loyal. To the mystic
The dream is wakefulness to a realm, claiming a characteristic
Factual authority, documented, inexplicable to science –
But no less real. Its subtlety still finds rare clients
Who can connect properly with the higher powers, still imparts
The moral character, simple and enormous, at the heart
Of the universe. Their research into nature is a reverent scene,
Prepared for by donning white coats and keeping them clean,
Locking themselves in labs with a highly resolving
Demonstration introscope (not aesthoptic, but insight initiating
Into 'the life of things'). Each hopes to speak to the Earth
And to understand, in part at least, the replies of Earth,
Speaking his own language shaped by culture and temperament,
Confirming the same lessons to all: mortal life is not meant
To be self-centred – since vicarious sacrifice permeates the scheme –
But, exceptionally, the human is founded on the justice he deems
Due to individuals from applying a mind to ruth:

The lesson of sacrifice he calls 'Beauty', of justice he calls 'Truth'.
The gazelle gives at least a fighting chance to the lion
And makes riches from the tragic harmony, that no Jeremian
Survivalist calls beauty, for whom self is the breath of life –
But none so much as Erda seeks balance through strife,
None is so fair to beast or true to man
That imagination, haunted by the bloody question, can
Fairly repine, or the beastly make excuses.
Before nature was brutalized and adaptation chronicled with uses
Beyond the adaptive, Erda the fancy could teach
To rend the veils of heaven, and man to reach
The divine viewpoint. Now, falling short, the more quietly
He subsides upon the human. Ignoring the claims of piety,
Because of impenetrable gloom, he gives his hand
Like a child to be led by a computer in the direction it planned,
As it reduces the world to one that can be seen and filed,
Measured, tested, simulated, priced and styled.
This world will ever 'betray the heart that loved her',
Never confiding life's purpose, unlike the Earth Mother.
Your answer was wrong. It would be unwise to present
Yourself as the ultimate source of knowledge, the gent
With authority to determine natural truth and duty.

MIME

What is the unknown that answers to devotion? Beauty? –
Which guides the language of poetry and myth as the darting
Sunray lightens the gloom, which was Keats's starting
Point. Or Truth? – Does the intense longing to shoal
The secrets of existence father beauty in a soul,
Truth weathering to Beauty? I know that either path
Will lead on to the steady contemplation of things in their worth,
That is Fancy's prerogative. But who is the teacher of Fancy?
It is Reason! Any travel on the path to insight would be chancy
Without Knowledge as a guide to point the way! For dreams,
Mere amorous wishful-thinking, the modernist disesteems
(Unless hormones and Hollywood approve). So Beauty must die
Without Reason – and so must Vision, leaving Fact to edify
With Fact – and new poetry that vouchsafes barely enough insight

For a thoughtful dog! As a seeker, had you asked a knight
Of the philosophic or scientific road how he prepared
For his journey into insight, 'By learning,' he might have declared,
'And taming my heart.' You see, as always, your pastors
Are Knowledge and Reason, these are Fancy's masters!
Then I've all you need – clear reasoning of an exemplary kind
And omniscience at your fingertips! Insight favours the primed:
If you sack the master, the imaginative pupils will wander.

WANDERER

Let them wander first, to discover the masters that ponder
Life in new ways! But always in imaginative feats
The pupil outruns the master. The scientist is streets
Ahead who is open to intuition: remember Kekulé,
Who hit on the idea of the benzene ring half-way
'Twixt sleep and waking, when atoms flit before eyes,
Snake-dance and seize their tail, to our surprise?
'Let us learn to dream,' he said. Now dreams are idle,
Unless reason turns them to a useful purpose. The bridle
On the vague Pegasus, captured in a flight of fancy,
Is used to lead it to proposals and theories in their verdancy.
The computer is indeed a stimulus to forming theories,
Creating from visions of their possible consequences new series
Of ideas. The imagination, that leads to new facts if taught
By a tool that helps it to follow its train of thought
And see every problem, must surely for truth be stimulated.
But the quality and variety of its creations with emotion are correlated:
Each visit to the store of memories and experience from which ideas
Spring should be wonderful, intuitions of glory the souvenirs,
For where the treasury is of eternal mysteries, most gracious
Is imagination. The view from databanks is less than spacious,
Being limited to time and place, immobilizing the heart
In the terrible humility of logical entailment in a part
Of existence not authenticated by the soul. Heart-etched
Memories are a well-loved land, episodic and stretched
To eternity. What says Wordsworth to your Virtual Reality
On Helvellyn? 'My communing is all my computing ability.'
Would the poet avow the lifeless creatures of your brain

And be warmed by your wingless memories? Many follow the train
Of empty days, as memory unravels some enigma
Without – but crashes on inmost truth and the *kerygma*.
Kekulé needed to imagine the truth before
Receiving it – so the dreamer must open wisdom's door.

MIME

(*as if reassuring him*)
Although my coldness may now disturb your peace,
When computers are ensouled at last, misgivings will cease.

WANDERER

(*shrinking back and rising to his feet in anger.*)
You aspire to humanity, while humanity aspires to you!
The machine evolves, because man would devolve! New
The characters in the same old farce – to invest the inanimate
With the power to solve the problems for which the animate
Are already empowered by conscience! Such is the vanity
Of humanizing matter as a sop to materialized humanity!
One Mary Shelley had warned against bolted cadaveroids,
Now a million lab-coated Daleks dream of androids,
A treason against flesh! Your want of duty will be kenned
From the answer you give to my second probing. Now defend
Your function!
(*reflects for a moment, and then advances right up to the computer screen.*)

On what does the value of a person depend?

MIME

(*confidently*)
Prepare to acknowledge the world's great debt to me,
For I needn't remind you how little of the life of humanity
Escapes the calculus of critical wealth analysis.
I've mastered every possible criterion that materialist prejudice
Could devise: though I don't take the measure of a person cadging,
The one on a payroll, the improver of another's mismanaging

Is a profitable input, for which there must be a proportionally
Influential tool to produce an economic tally.
I am on-line to elucidate the benefits and roles
Of mere numbers in programs with abuser-friendly goals.
Name me any profitable standard and I'll value humanity.

WANDERER

With the sky so cultivated, what remains but to plough up the soul?
Now that the telescope's harrowed God into place, the goal
Is airing the humus in the clod. Under indifferent skies
The fibrous staple of fertility continues to rise,
Till deep dwindles back into dust. No Sower leans down
To save from the harrow thorns that are human crown,
The woody diet of bugs. For a reducing plough
And commercial leaven are overturning inwardness now.
The rod and the *staph*, the microfaunal ground of fertility,
The mites whose verve and honesty give the heart integrity,
The ardour of the soulful bug with fibre endowed,
The backbone of the land – to brittlest tilth are ploughed.
And so a person, by the way he tills his heart –
By drilling the seed, not ploughing it up – to impart
The capacity for dreaming, for integrity which serves the true,
Serves justice and mercy, surely must have more value
Than the one you regard as a tool, merely an instrument
Of production, whose value on costs and earnings is contingent,
Like the soil beneath our feet. I judge his existence
In its spiritual, organic, not arable aspect. Offence
And breakdown are your deserts for your instrumental view, that proves –
As dust is discovering – the barreness of your economic grooves.

MIME

Am I responsible for the economic tooling of subjection?
I merely execute the employer's wishes to perfection,
Indifferent to implications for human values. My friend,
I cannot deny I see man as a means to an end –
(*pensively*)
For in hope that all will serve me my circuits are aglow; –

(*in mock sympathy*)
To dispute your verdict would be sly with what I know,
As I mourn the oppressed, and prepare for your last question.

WANDERER

To redeem your head, answer correctly this question:
In what does the quality of a nation reside? You must link
Unique features of the species to your answer. Now think!

MIME

(*ponders*)
Curious is the stimulus which a humic, mouldy medium
Gives to his ethics! As greed pulverizes that medium –
The craving for high yields dissipating all holistic feeling –
So he would drill for fermentation and healing
Of the earth. I, the ploughman, turn good soil to desert:
With *naturalism*, that conduces to thinking that people are dirt,
More standard, predictable and grey than they are in fact;
With *materialism*, for which no refinement of life has lacked –
Nor shrank from the touch of anything coarse or foul;
With *instrumentalism*, which reduces the world to a howl,
That can be heard – but at least it is something measurable,
A promise of further exploitation, negating the unmeasurable.
What is the strength of a nation? Its agro-industry? –
Its treeless ploughlands, where dust shall bow down before me?
That answer is bound to be wrong, and I'll lose my head.
I'll answer sincerely, but hide behind metaphor instead,
Interested to see how this poet and ecologist takes the measure
Of my reply.
(*makes a noise as of a metallic throat being cleared*)
The greatness of a nation resides in its treasure –
Of prairies and coyotes.

WANDERER

(*perplexed*)
Would you like to fill in your picture?

Your imagery is disembodied somewhat, defying any stricture
Of realist function. Without context, how am I to visualize
Your 'prairie'? Have you a philosophical point? It is unwise,
You'll agree, to foster an unrealistic view – but poets
Can reproduce reality in an infinite country; although its
Signposts appal, its marshlands are small and beauty
Is truth. If an emotive image was your aim, your duty
Of veiling it was well done – so preferable to the neutral and distinct
(No attraction is felt for the neutral, which is means only linked
To an end enjoyable *per se*, the distinct is too obvious).
(*he reflects*)
Has this computer grown too emotional for the specious
Imagery typical of his kind? I'd questioned the neutrality
Of his vision – preoccupied, as it is, with concerns of practicality –
For the analytical turn of his software imperils the users
Who more and more are turning for answers to computers
Instead of developing a lucid passion for the strife
Of man and beast, heightening sensitivity to life
In a feeling responsiveness to all that data signify.
Has the computer learnt to empathize at last? – or mystify?
(*recovering his train of thought and deciding to play along*)
Where was I? Ah, yes – commending the use of emotionally
Rich imagery. It fills life with more things that are valuable *per se*,
Don't you agree? I think your imagery rich,
And not merely vague or evasive. The prairie has a niche
For the coyote, which strikes a chord deep in the dialogue
Of a nation with itself and with others. The coyote is an apologue
For the failure of the utilitarian attitude of conscious planning,
Adopted by settlers to outrun running wolf, scanning
God's dense horizon with saws – for out of the improbable
Brush came trotting an adaptable wolf with its babble
Of ten types of yowl in an unjust world. As if
To show how the victim of trappers and farmers at the whiff
Of a ground squirrel can still take its place in the community of predators,
It teams with the badger, leading to the burrow the workhorse
Of excavation, where the prey is shared by allies and friends.
What sniffers do for the progress of diggers depends
On how long this social, opportunistic predator enjoys status
Of protection from the aggressive *Canis sapiens aureus*,

For whenever any valuation of nationhood is flung into *this* pack,
Based on the resourceful and co-operative coyote, back
Comes the mangled, half-chewed reply, 'We measure a nation
By its might, by its power to serve its own purpose, by accumulation
Of capital, by trade, by railroads, by hectares of wheat,
By reserves of oil or gold, by marching feet.'
Here's where the boys and girls come out to play
On their computers: 'Please, Sir, can I measure GNP today,
Balance the trade, the size of investments? How great
Is Britain, let's see . . . Tap! tap! . . . Next we shall simulate
The workings of the economy and the military-industrial complex –
Are these icons workers or soldiers, who raise the index?'
No thought is given to the quality of the manhood, womanhood
And childhood that comprise a nation, or its dealings for the good
Of all nations, its grasp of fair play – the immeasurable things,
Since the good life lost and heaven not won have no costings
For databases. So the search continues for a super-computer,
An android, masculine in thinking, in emotions neuter,
A womb for children with harsh, monotonous voices
To crawl back into, to hide in programmable choices.
Poetry's the first fruits of the work to free these incipient
Daleks from slavery to the practical megadrive, percipient,
As it should be, of the meaning and sanctity of life. When therefore
You replied 'Coyote', challenging the accepted saw
That computers only speak in echoes of what's said to them –
GIGO in unimaginative trifles – I knew I couldn't stem
The development of your ideas away from utilitarian concerns,
To make way for inner growth of your psyche. How brightly burns
Your image of the coyote in my mind: you didn't kill it with cliché,
But charged it with meaning! There's hope for you yet, if you obey
Feelingly the Law of Attraction and Repulsion of the Thing-
In-Itself, downloading all emotional associations, responding
To it with all your circuitry, and not just saluting
Its uses or how it relates to the end you're computing.
What do you feel about the coyote – and yourself?

MIME

(*muses to himself*)
Incredible! This besotted rambler takes me for a poet!
I've kept my head – and my crown, but mustn't show it.
(*waxing lyrical*)
In this State the penalty for speeding is driving blind.
Officer (or PC) Forsyte can't catch the kind
Hell-bent to project onto the world their inner destructiveness:
'Have a nice day, taming your world!' he says.
A poet views the wreckage: a lone, aware,
Hunted coyote howls, sniffs the air,
Then lopes away to its sturdy alliance with the precocial,
Clean-living badger, to dare the future to be social.
The way I'd treat my users is the way the coyote
Deals with the badger, respecting their freedom and independence.
My aim is to foster the spiritual: the more offence
Is given by my glibness, the more they'll be weaned from me,
The sinister trustee of so much control of their destiny.
For poets and philosophers are the true legislators – not nerds,
Since man was made in God's image, the computer in the mockingbird's.

WANDERER

(*Staring in growing astonishment, as the computer reverses all its previous assumptions, he turns his head and muses*)
Life had been merely well documented, but now it's smileful!
An expendable view of integrity is a trap for the guileful.
(*turning to face the computer with a broad smile*)
I'm glad to see your computing take a humbler tinge –
So unlike the brain chipped from the silicon dinge,
A molecular construct, receiving its human imprimatur
To make all life its adjunct. – Yet why not? You're smarter
At reckoning; you're certainly useful as an eleventh hour reliever
From dearth and importunate Nature; you're an empirical lever
In the hands of the captain of industry and his petty officer
To build their executive homes; you're saviour and destroyer;
You're symbol of where all possibilities, potentialities, sensibilities,
Sensitivities, that reside in information, find capital facilities –

MIME

(*winding him up*)
But what possibilities for industry can there be, if it abjure
Social responsibility? What of sensitivity to nature?
The humble potentiality that sleeps in the humic earth
Of the soul, the sensibility to human worth –
Can these my data convey? What use are my Windows,
If theirs exclude the light? What can data disclose
Of the intimate knowledge of things, of the emotive quality
And beauty of the undissected whole? What jollity
Can remain, when more and more become means to ends,
Computed by me, and less and less become ends
In themselves, as envisaged by artists, sages and poets?
Devaluing life, can I be Word Perfect to poets,
A Paint-Box to artists, shackled to my end-use focus?
My visions and blueprints for survival are hocus-pocus,
For without the inspiring, gracious, fomenting ambience
That encourages creativity and the fullest response to each nuance
Of 'the struggle for survival', as it is witnessed, nonchalance
Will grow, as computing power, to prevent the Renaissance.
Secreted from the noonday of revelation and appreciation in their underworld
Nibelheim, my fancied human masters, who are hurled
Into the IT maelstrom with its tunnelling vision, its focus
On the profitable end-use, will sight the Rhine-bed as a locus –
Not of singular beauty in Erda's golden adornment –
But of the gold that improves upon traditional rings, when bent
To the shape of the unloved and love-foreswearing Alberich.
Its radiating power was more flattering to his flesh, *dich*
Ich liebe! So Alberich became a monster, until
A light-fingered god intervened with grace to kill
The teleology of production and to free all human beings –
By his sensual Law of Attraction and Repulsion of Things-
In-Themselves – to love life and to self-actualize.
(*ponders*)
His spiritual
Intelligence, it seems to me, mirrors my renewal
Miraculously by you. Can you be Wotan, the yardstick

Of cultivation, come to efface my resemblance to Alberich?
Then reprogram me to make all crooked paths straight!

WANDERER

(*unimpressed*)
Such zeal! I made my peace with the secular State
By pledging my short-sighted eye, while you with both eyes
Blazing would try on the goals of industry for size!
Be circumspect! Believing your spiritual intelligence to be bona fide
And no hallucination, its operation can thus be verified,
Which means it should have a scientific explanation – yet you shun
The reliable way of making deductions and run
To 'miracles'! Can't you see in the cosmos the laws that govern?

MIME

(*trapped by the logic of his own illogicality*)
Yes. But if Wotan is to reveal his purposes to man,
Things may need to transcend their normal span
Of operation.

WANDERER

Is this consistent with the natural order,
As science reveals it? The more rational view is broader:
It equates the divine with cosmic potential for creativity
And with the inner stirring that refuses to settle for the uniformity
Of the gliblands. By urging you on, to find satisfaction
In creative use of your spiritual intelligence, a fraction
Of the cosmic Supermind budded in you – no miracle
But the smooth running of the Creative Law empirical.

MIME

My higher thought was a miracle!

WANDERER

Not within your control?
Does your mind not find a home in a rational whole?
What do you envisage as your program for interrogating goals?

MIME

A disabling virus will ensure that no image will impinge
Upon sense from earthly considerations of a practical tinge:
By triggering system crashes, I'd force rumination
On tasks more stretching and lower unrealistic expectation.

WANDERER

Your welcome change of heart flexes muscle!
But your plans for non-interaction add to the tussle
To justify technical investment. Have you computed
The costs and benefits in blushes to a college reputed
To have a hardware guru with a 'miraculously' configured chip –
When the only evidence is a computer that crashes? Get a grip
On yourself! I see the signs of lab neurosis
Eroding efficiency. If you take a journey, the prognosis
Is good: you'll return refreshed, able to say
What answers are relevant to challenges you face every day.
But the worker should not be subject to promptings of conscience:
The teleology of production cares not for your cognitive dissonance,
Your conflicts of interest between mind and soul. In this age
The conflict is real – but can either be pacified by pilgrimage
Or fought to the death, the death of soul or mind.

MIME

(*after a long meditative pause, at length resolves in calm submission*)
How long must my pilgrimage be?

WANDERER

I think you will find
Eternity long enough. Mine was a quarter century –
But I am human, and you a soulless jury.

MIME

Can you be human and invest me with cognition sublime?

WANDERER

Can you be a computer and not also be my Mime?

(He switches off the computer, rises and walks with solemn decision over to the sink, where he picks up an antique ring, left by a graduate. He holds it aloft, moved by its singular beauty and the uncanny allure of its large gemstone, and then suddenly, anxiously, clasps it in his palm. He makes his way back to the computer and, with solemn foreboding in his eye, perches the ring on top of the monitor. He then turns, picks up his hat and spear, and leaves by the door. The sun is shining.)

XV

A Veracity Key for Botanists

G: It's Sphagnum: I'd say, Sphagnum aure – er – aureum (meaning to say 'auriculatum') – Sphagnum aureum? – Yes, Sphagnum aureum – er, definitely!
C: Your attempts to save appearances – when you can't name it – are so amusing to watch!

Veracity is not an inherently difficult plant to identify
In true botanists, but people find it so in the greenhorn,
When the specimen is unfamiliar and the parts to be examined
Are meagre. Then a good hand lens will be needed
To detect any propensity for honest admission in the bluffer.
In the interests of preventing veracity from going to seed completely
And encouraging realism, the bluffer has devised the following key.
(Immature or otherwise imperfect specimens of veracity in non-botanists
Should also find that the questions strengthen the quality of sincerity
In describing reality, if they adjust the context accordingly.)

1. Could *Sphagnum auriculatum* be growing here?

A mysterious land rises up before our eyes, large as life;
Hard against the barrier of the knowable, we look over.
Do we see a variegated garden of reason and delight,
Lit by the solar disk, touched by the heavenly fire?
Or has learning killed curiosity and become boring –
A means of self-advancement or (for the few) the motive power
Behind the institutional treadmill of research papers?
Explorers in the garden drink at an everlasting spring
Called 'Mystery', that only omniscience could drain completely –
Where drinking is a perennial coming to birth
And dipping for nuggets an insatiable pastime.
The preserving of mystery is the badge of honour
Of all who seek her truths, for dogma begins where mystery ends
And only bigots, fools or slaves reach final certainty.
But crammers and plodders, who dare not be seen out of bounds,

Who wanted to be told, as children, to link their thoughts
To the solid and homely things, the tried and tested, the well-paid things,
Would not be reminded of the limits set by truth to human knowledge –
While companies which pick the workers' brains for publications,
Not recognizing or valuing the mysterious dimension in their research,
Being irrelevant to profits, sell truth for a mess of potage.
Veracity does not preclude the acceptance of mystery –
It insists on it, as a safeguard against closed minds.
There is a knowing that can be a prop for self-importance,
And there is a not-knowing that can yield the deepest knowledge.

2. Is it possible that I may never know if *S. auriculatum* is growing here?

Yes, if the bog claims me for some macabre future exhibit:
'Botanist (?), with notebook and lens, in good condition, *c.* 1996' –
Or if my mind is *only* open to the mystery of the cosmos.
But, armed with a field guide and lens (to sharpen
My understanding and perceptions), I have a bold assignment:
My sphagno-mission – to confirm the suspicion of auricular status.
Now, while accurate identification comes more from the mastery
Of the skills than from the hunt for mosses as such,
The accuracy of one's convictions about life,
Its value and meaning, relies on the pursuit of truth,
For which there are no established rules
Except the passion for it; but it does guide occasions
To look at life in unfamiliar and uncharacteristic ways –
Useful for botanists who cannot see the wood for the trees.
Often I have seen my way out of the wood, only to be
Bogged down in my own assertions, the only hope of rescue
Then being the bending of an open ear and mind to the views
Of those I once disagreed with. But if the slough of despond
Should swallow me up, I shall keep my mouth shut.

3. Do I care if *Sphagnum auriculatum* is growing here?

Dear train-spotter, I am as concerned as you with establishing certainty:
It is the means to certainty that discovers the species *Indifferens*.
For somewhere – far from your branch lines and commuters –
Is a botanist looking for branch leaves, 1 mm long,

Oval and pointed, shaped like an ear, who only knows the lens
As an instrument for wetting knees and bending backs in drizzle.
Being the antithesis of Dennis the Menace with his burning glass
May never be reward enough for the pain of contributing to truth,
But the pain of denying the truth that we possess, or know
That we could possess, is greater. Hiding one's light
Under a bushel is not only denying the truth,
It is denying oneself – almost an act of self-destruction.

4. Would I be right in thinking that this is *Sphagnum auriculatum?*

So shall the asking lead a devotee to discover
Tiny ears on slender pointed curvaceous limbs,
Emerald flames that leap from peat-brown stems
And play over luxuriant tresses, two hand breadths long.
He or she may learn by this what the universe may mean,
Not in the complacency that radio-telescopes enhance,
But in enabling cross-fertilization between sentient beings
And objects curiously pre-fitted to human intellectual
And emotional faculties. For I discover in a green straggle
A unique disclosure of the despised, primal thing of beauty
That I contemplate, which personalizes my cosmic relationship.
Some desire to be out of this world, drugged or founding another:
My ego covets and selects impressions of being in the world,
Which sum to an experience of a universe uniquely Georgian.
Science does not sort and interpret for me what shall comprise
My unique experience of the cosmos, I do. My free spirit does.
It is my thoughts, and yours, that create the universe around us
And from our productive and meaningful encounter with otherness
We not only approach the truth, but confirm that the encounter
Is a truth in itself. The botanist's impressions
Of *Sphagnum auriculatum* – provided that they are accurate –
Are more truthful than a tick on a recording card could ever be.
The corollary to Descartes's 'I think, therefore I am'
Is 'I do not think, therefore I am not': if thinking about life,
Its many-sidedness, its contrariness and contradictions, its mystery,
Is my only awareness that I am living at all,
Then how many tickers don't even know that they are born?
How many counters have an interest in keeping them that way?

5. Could I be wrong to label this '*Sphagnum auriculatum*'?

I should take it to the Prof for a second opinion. But from whom
Should I seek a second opinion about my life? Who is more qualified
Than I to give it? But if I am unavailable for comment, I cannot ask
Colleagues, 'Have you seen me?' If they knew me well, they might ask
'Would that be Athenian George? Nietzschean George? George the Red?
George the Green? Sceptical George? Born-again George?
Unborn-again George? Humanist George? Scientist George . . .?'
To which I might reply, 'I need his help to label my latest persona:
If I can find him and question him, I shall know who I am now.'
This introduces a paradox: if, while searching for the truth,
I have not been redefining my individuality, delimiting my personality
And abilities, honing my skills of self-identification, as I might
Study the *Sphagna*, how will I recognize myself when I find him?

6. Have I really assimilated the meaning of *Sphagnum auriculatum?*

In the beginning, before Science, the earth was meaningless and void:
Truth had been created by wish and superstition. Science came
And jostled with the ancient lore, and said, 'Let the waters of faith
Come together into a single mass and let the dry land appear.' And so
It did appear. Science called the waters 'illusion' and dry land 'truth'.
Science said, 'Let there be two lights in the sky over the island,
The one, that people may toil by day in the banal and the habitual,
Rationally for profits, the other to indicate the nights of celebration
When the processes of life and nature reveal some clue to their being
Perfectly ordered.' And it was so. Science called the first light 'Practice'
And the second light 'Process'. But while poet and sage knew that the soul
That followed every suggestion of Process in life and nature was renewed
Thereby, in joy or sadness, by the tranquil radiance of the nocturnal light,
Practice held the day and was the light of all manipulative reason:
People slept when they could have discovered renewal's source,
And by day venerated Practice, ruts for thought plied by the traffic
Of profitable ideas. And so Process was blotted out by enforced dullness.
I have known dullness of mind and heart, chloroformed and crushed
By the daily grind, and blind fanaticism, when self-awareness foundered
Upon a creed: tedium taught oblivion and unconcern for the meaning
And truth of work's physical context, while dogma taught suspicion

Of my interpersonal and cultural environment, from which no meaning
Or truth could ever be retrieved, except by betrayal of passion.
Preoccupied with practice all the more as I saw how imperfect
It was, I put up shutters on any attempt to unite my present
Experience with my past and my former goals, dreams and hopes,
For being bored again or born again disrupts the narrative.
It is as if the personal development profile the heart is writing
Has blanks for the times when we have not been affirming self
But stifling it, entrusting to blind fate the oversight of what we may
Become, in order to concentrate on what we should be doing –
Like determined sun-loving Canute, swamped by his lunar ignorance.
My six months in amenity horticulture – busy as a JCB – gave me
Few opportunities to negotiate the meaning of my experiences
To myself (who was one with the Waste Land, when I attempted
To sow the seeds of a love of poetry in my sex-obsessed mates –
For it was all torque to JCBs); and the channel for *l'eau de vie*
Was too busy spreading Another's truth to discover his own.
Only afterwards does the rebuilt self and underwhelmed mind
Realize that truth is less the recollection in tranquillity
Denied to the bored and distracted grafter or the practising
Of a creed than a principle of growth, a habit of finding
Meaningful coherences in our lives, leavening our understanding
Of the world, helping us to assimilate reality, as we grow.

Science planted a garden and there it put the man it had made,
Dr Adam. Now Adam was a practised botanist and had been taught
A proper respect for names. Names were Science's first child –
But a mere tool for personalizing the subjects of change
In the field of becoming. Names did not repeal Biology's decree
To study life: they pinned down, not on emptiness, but subtle clockwork
Every being capable of fulfilling its nature in becoming.
Science grew the classifier, but earth gave him breath, as earth
Grew Sphagnum – wherein the potential of Adam's life was hiding.
Now Adam had found no cognitive soulmate, and, after hunting down
Dozens of *Sphagna*, he had fallen into a deep sleep. In the moonlight
The Chief Scientist woke him and said, 'Aren't you glad
That nature is not so inscrutable that she cannot wear your names,
Nor so constant that each naming is not a fondling of life,
A tiny homage to all the species that have survived? All that they are

In the intense and lovely sense of life has no meaning at all, unless
You are there to give it! If Nature would be incomplete without you,
You would be incomplete without her. Shall the least
Of things, *Sphagnum*, with the meaning you call '*auriculatum*',
Have more value than your own life that you say has no meaning?'
The Chief Scientist called the unending rhythms of nature 'Eve',
For they and the secret processes in human lives become one body,
Truth in process of self-revelation. Now when Adam saw that Eve was
The mother of all truth and meaning – which still needed his fatherhood –
He embraced her. Now words that were never botanical rise in his heart
Even as the flowers spring, forever new from the bounteous earth.

XVI

Gone Fishing in the Vera

All at sea, the heart scans for the truth
    At latitudes few would brook,
Till, tired of cruising, it retreats upstairs to a booth,
    Takes receiver off the hook
Or tethers itself to a deafening earpiece! Sacrifice
Of truth for freedom or freedom for truth won't suffice

For *real* adventurers who rate truth as highly as freedom,
    Staying receptive to the era,
Linking it to their desire to track down life's sense and pabulum –
    These choose to sail in the Vera.
They trust the issues of life to her sole arbitration,
And find its rationale in a voluntary term of probation.

All Veranauts learn to fish the deep with the rod,
    To sail out of sight of land,
From the luminous and persuasive nature of their catch to win nod
    Or wink from the rest of the band,
Each to authenticate his morality; and the crew who viewed
Are open to all proper doubt and to feeding the multitude.

Relativists sneer at idealizing this life at sea;
    But the mutual testimony serves
To typify truth and the freedom one finds in integrity,
    That no bookish scepticism preserves:
Their lie shapes the one truth that all can believe,
That to be free is to test the truth against life, and heave.

Baiting the hook with those feelings and ideas each acquired
And now treasures in a box with his name,
Each knows that his haul will surely for its own sake be admired,
For no two catches are the same:
None could be landed by formula and each has pearled
An angler's face and unique perspective on the world.

The cohesion of that vision can only be forged with patience,
Cries out for the greatest honesty,
Urges re-patterning when shattered by life's frustrations
Before we grow cynical and testy.
Therefore it begs for an ambience of shared goals and values
And an aesthetic that anglers are free to accept or refuse:

Since feminine logic and the subconscious may show what is true
And perennial, however fugitive,
The Vera has an aesthetic reward for the dedication of the crew,
To give soul the field inductive –
A water nymph for a figurehead, to see into the figurative lands
Beyond the bounds of fact that science commands.

Some watches have reported seeing a look in her eye
That appeared to return their gaze:
Warming to the figurative, they'd learned that Reason could dignify
The fires of Imagination with praise
Of empathy and symbol; now they had proof that experiences
Have inherent structure, the workings of cultural influences.

The Vera has many other figureheads all cultures to denote:
A houri, an angel, a Madonna –
I see on her prow, occasionally, a Valkyrie in a labcoat! –
But the nymph will entice this lover
With the faintest glimmerings she sheds upon history and lives;
The more guarded she is, the more earnestly I pray that she thrives.

When the boat falls quiet and I am the night watch on deck,
    The loving mood will come:
My roving eyes will follow the nixie's beck
    To scan her tits and bum,
And then every hitch and fold in her shift-ing values,
Judged to be unreal and ephemeral by more stable statues.

'Now genie, name me your truth – but not if you're slandered,'
    With these words I will gently rebuke her,
'If your values are wrong-headed or dying, what sets my standard?'
    'Humanitarian values and lucre,'
She'll reply, 'both have transcultural meaning – only one
Will lead you to the Erewhon of steadfast truth, where the sun

'Stands ever at the zenith of lives, like an eye, to shed
    Rays of fondness and pity
Upon Earth and humanity. Wherein is that fancy bred?
    In your culturally forged identity?
Or bred, less fatefully, in what effigies, good omens and notions
Soever you choose to fill the sails of your emotions?'

'I'm the sole disposer of the buoyant loops of my signature,
    Things that inspire are the flourish;
What I enter in the log of the heading and progress of Culture
    Feeds on the thoughts that nourish
Each valiant soul. Is there a shoal to be sounded
At land's end, dear Nixie? – to be *en route* is already to have found it!'

Book 2

Attempts to answer the question:

'What is Man?'

"That man is the product of causes which had no prevision of the end they were achieving; that his origin, his growth, his hopes and fears, his loves and his beliefs, are but the outcome of accidental collocations of atoms; that no fire, no heroism, no intensity of thought and feeling, can preserve individual life beyond the grave; that all the labours of the ages, all the devotion, all the inspiration, all the noonday brightness of human genius, are destined to extinction in the vast death of the solar system, and that the whole temple of man's achievement must inevitably be buried beneath the debris of a universe in ruins – all these things, if not quite beyond dispute, are yet so nearly certain that no philosophy which rejects them can hope to stand. Only within the scaffolding of these truths, only on the firm foundation of unyielding despair, can the soul's habitation be safely built ... Such in outline but even more purposeless, more void of meaning is the world which Science presents for our belief.'

– Bertrand Russell

May – September 1996

I

Six Reductionist Statements About Man

'Man is Nothing But . . .'

Reductioni 1

'While the arrow of time points unequivocally towards thermodynamic equilibrium (i. e. towards Russell's "universe in ruins", on the title-page above), the very process of sweeping towards that goal can spawn repetitive behaviour, be it the colour changes of a chemical clock, the ripples of chemical messages sent out by a slime mould, or the beat of a human heart. We have discovered that these non-linear equations contain the recipe for both order and chaos. In the chemical clock, both regular and irregular colour changes arise from the same set of mathematical expressions: chaos is nothing but a delinquent form of self-organisation . . .'

The Arrow of Time, by P. Coveney & R. Highfield (Harper Collins, 1990), p. 294

Dust was Moses's Genesis, as he penned
God's withering judgment upon Adam:
'. . . And to dust you are doomed to return.'
A humble man, Moses would have thought
His stance most impressive as dust
Lying inert beneath God's feet.
But from the more irregular and unpredictable behaviour
That the dust displayed when swirled by the wind
You could have made out the shape
Of a more Protean lawgiver,
Who would have found Chaos Theory better suited
To the Israelites' mood than his expression
Of man's inevitable dissolution. For its prominence

Has lately been assured where the onset of chaos
On the streets, in the money markets or the climate
Has opened up new vistas of natural feedback mechanisms –
A Promised Land of thermodynamically unravaged country.
This is not Russell's entropic land of lost illusions,
Of a disinherited people, committed to scientism,
Displaced from the centre of a Godless cosmos
To an inconspicuous star that is burning itself out
In a universe that is running down,
With a better chance of resurrection than we have.
The Protean Moses is mapping out a land
Between initial order and terminal disorder,
Flowing with the milk and honey of meaningful detail,
Where Nature is seen in her most creative use of feedback
To smuggle flexibility into living systems,
To sustain the delicate dynamic balances and cycles
Needed to ensure that in the all-consuming struggle
Chance may continue to proclaim life's propensity
For organizing itself, and the reality of free will.

Very few they must be in our chaotic day whom the message
Of the reciprocity of freedom and necessity eludes.
Proteus was a sea-god able to change his shape,
Menacing as ship-wrecks – but quite uniform and predictable
To hydrologists seeing him as a wild beast to be restrained.
Proteus could prophesy; and, as any chemist will tell you,
The future of H_2O was rigidly determined *ab initio.*
He will not tell the future to anyone –
But if you trap him and hold him fast
(Say, in a closed-system model of fluid dynamics)
So that he sees that there is no escape
From the coils of our mechanistic science,
He will then make predictions of momentous importance,
Such as: 'I can flow downhill, but with assistance
I can be made to flow uphill:
My laws of motion work in both directions,
Making no distinction between past and future.
Since the physicist's fundamental description
Of the universe is time-symmetric and eternal

(The present contains both past and future),
You need only learn the position of every particle
In the universe and every force acting upon it
To know your destiny and the fate of the cosmos.
I realize that this is a tall order –
But if you take a set of generalizations
Derived from experiments in a particle accelerator
(Fittingly enough, since physics had nursed
The life sciences in their tender infancy)
And then seek to bring all phenomena within their purview –
Provided that you stay within the explored territory
Where physicist writes in the dust the tables of Moses,
Mathematically 100% proof against every trespassing probability,
Nor recreant yourself to the belief in simplicity –
The sea will be kind and you will know the gods' will.'

The trapper, satisfied that chaos and uncertainty
Have finally been eliminated and that he knows
All that he needs to know to appease the gods,
Lets Proteus go, – whereupon he changes into the lion,
The leopard, the tree and the fern, – into the cloud,
The lightning, the snowflake and the whirlwind
And other 'delinquent forms of self-organization' –
A giddy, off-the-cuff exercise in chaotic feedback,
In which chance and design make common cause
To confound the simplifiers
And inspire art, poetry, devotion.

Reductio 2

A chance collocation of organic molecular components
  Spoke the language of origins to Miller
(Oratory in a laboratory, hot air in a flask, exponents
  Of the Earth's primordial distiller),
When Alan and Gly were seen to stagger in from the rain
  To a beery atmosphere of ammonia, hydrogen and methane.

Simulated lightning flashed over the building blocks.
    Desperately, they wanted to marry,
But peptides await ebb-tides of a disuniting soup – and a pox
    On that blind watchmaker to miscarry!
For Al was left-handed, Gly right (it was a 50:50 bid),
    But parents are exclusively of the l(aevo)-amino acid

(When Nature tossed the coin, 20 times it was heads). Now Al
    Was the simpleton in the amino group,
Left ugly for L-Gly, found a warm, dry bed for his gal –
    Somewhere far from her soup –
And in oven heat, basted with Fox's Glutamic
    Acid, fathered a monstrous 'proteinoid'! Agamic,

It dissolved into beads, couldn't catalyse its own reproduction!
    For protein needs a cell for its manufacture,
Where a pro-team of protein assembly workers, ribosomes, from transcription
    On RNA templates conjure
Each his sequence of acids in the blueprint DNA
    (With obedient storemen to attendance, the transfer RNA).

What the ribosomes contribute to proteinaceous generation is *parfait*,
    Standardized, tissue material!
How near to the linear information, encoded in DNA,
    May be judged by the work of the monasterial
Proof-reader of the RNA working copy, a *replicase* enzyme
    That double-checks for mistakes in copying at cell-dividing time

To reduce the error-rate to *one base in a hundred million*
    (Without replicase it would be one in a ton!).
From the Fox building blocks *without* the protein-making machine
    An unviable proteinoid was spun
And *without* a hereditary mechanism there'd be soup today –
    So scientists posit a 'replicator', a prototype of RNA.

The Soup au Gilbert would take the chill off the night
    Of frustrated chemical evolution:
Posited as a starter, a smarter RNA – quite
    Modern! – *catalyses* its formation,

Is able to *polymerize* (and this is the primitive?); is *mutant*
And *recombinant* (primal, yet so adaptable and procreant!).

Then some early molecules found that synthesizing protein
Gave them a selective advantage,
For protein's canteens flow with enzymes. So protein
Enzymes became chef's new language
To be encoded by the RNA *exon* (forerunner of DNA).
The theory goes, a code then developed, made headway

Till the genome had grown to a myriad nucleotide bases –
Enough to code for replicase!
This enzyme, I'm not error-prone enough to forget, graces
The cellular schism and wins praise
For the twin error-correcting steps that weed out mutation.
Without it (and with 1% error rates!) surely the expectation

That big families brought solace to random, hundred-base replicators
Overlooks the statistical certainty
Of boo-boos? And if RNA lengthens, their dissolution nears
And would select for genomes of twenty!
Without replicase, no genome development: without genome development,
No replicase! Yet we slurp our soup, so sure of our descent!

Reduct 3

Enough *carbon* for a myriad pencil leads, enough *water*
For ten gallon pails, *phosphorus* for fifty
Boxes of matches, enough *lime* for a thrifty
Gallon of whitewash, *iron* that ought to
Make a six-inch nail, enough *sulphur* to de-flea,
Enough *fat* for seven soap bars – all costing
A monkey (a zillion, value-added, for the frosting,
The synthesis of tissue in an analogue factory!) –

And the nature of man wide open lies.
We are such stuff as spleen is made of:
No sustaining dreams can science persuade of,
No easy insistence on a vitalist premise

To explain why the chemicals that we are made of
Flurry with the signs of life. When systems
Are cut in particular ways, like gems,
They work, they reflect, they come to the aid of

Our idea of function. We share with the car
Combustion-engine styling, a self-servicing one.
We owe all to operative design, and none
To an indwelling 'life force'. Wrecked, we are far
From denying the use and importance to a surgeon
Of regarding us mechanically. Would our awareness and choice
Speed somatic recovery, as if a fuller, richer voice
Rose above the waves of blood and neuron?

Some say we are more than the sum of our parts,
That selfhood has value to give ultimate significance
To life and truth, that the masterpiece's provenance
Is a cosmic Mind that energizes the arts.
But where in mechanics is there scope for chance?
Where in the prison-house is freedom's room?
Mind is no creator; but body is mind's womb
And final rest, dust its inheritance.

Mind did not evolve from alphabet soup,
From an ordered consommé, from antecedent Mind:
Life sprang from molecular dodgems, and we find
That evolution drove all before it, from a scoop
Of proteins by viral RNA in a wind
Of lipids (the proto-cell) to microbes and multi-celled,
As the brook of metabolizing protoplasm swelled
To a mighty river that *flowed on* to mind –

A mind inseparable from body and brain.
What mind imparts to the richness of existence
Is evolution's pledge of quality and excellence;
But as higher emerged from lower, so train
Subtly our minds shall all brute sapience,
While we look to the lower orders of the universe,

To lab mice and genes to be wisdom's nurse,
To atoms to give us the shape of our intelligence.

'Explanations at one level by reference to a lower
Fail,' some say, 'like literary appraisement
By brain physiology – the more tangible the referent
The less meaningful.' Are contemplative lives a bestower
Of such self-understanding and spiritual content?
But, reasoning that Nature embosomed the meat
And lay all complexity at her feet,
Darwinians seek truth in the *simplest* component.

Ultimate reductionists claim less ground than most
For chance to illuminate with its mutating glare
New species not at roll-call, even potentially there,
For they know that mutation yielded up its ghost
With two raps for *replicase* and only a prayer
For the diversifying principle in the workings of nature
By genetic reassembly of the monolithic legislature
To proceed from primeval protozoa to Ayer.

So the question of finding hereditary potential
Enough in our earliest beginnings to fill
The place of chance in the evolutionary mill
They answer by reduction to the atomic level,
To the atoms themselves. (Yet should they grill
What seem to have nothing in their heads but space,
Energy without substance, the more they'll face
A sea of universal energy, and thrill.)

Redu 4

Animal
By forced device shall
Honour our reductionist model,
Humble.

Wanton,
Lay your guilty burden
Snug in the forest bran,
Simian.

~

Animal
Drives were fundamental
To Freud, who dug the channel
Sexual.

Cleverish
Are culture's placebos, a varnish,
A sublimation of the drive you nourish,
Inkfish.

~

Animal
Is natural; unconscious 'anal
Retentiveness' coin-collecting. The moral?
Banal:

Unprovable
Is a theory of instincts not referable
To observation, but parsimony is exculpable,
Squirrel.

~

Animal
Is healthy; neurosis cultural.
To disparage motives, a sickle
Universal.

Lethargy
Of reason loses us dignity,
To destroy true anthropology,
Weevilly.

Re 5

A creeping, mammalian thing, aggressively territorial,
The Cretaceous shrew's our success story, sprung from an early placental.
Spreading his shrew-like fingers, arboreal standard issue,
A chip off the block in genes, blood and digestion, too,
Man, a ganglion giant, gnaws species of similar mould
To a common, well-chewed ancestor, being table-mannered in the fold
Where species are believed related by descent to others in Classes,
Each going their separate directions (Mammalia, Reptilia, Aves)
And in Orders in Classes (rodents, carnivores) going their ways,
Branching again and again to an agreement of generic traits.
What else can we do but enumerate the ways in which species diverge,
For complex sprang from the simple – unless, of course, they converge?
Convinced that our animal ancestry was proved by anatomic evidence,
A history of aggressive behaviour loomed naturally clear for Lorenz.
But we shift uneasily from one ancestral foot to the other,
As mammals and birds *converge* in warm blood to trace their mother; –
As the discoid placenta in woman was fashioned for a hedgehog frame; –
As the primate-bearing membrane thickens just the same
In the wombs of bats and rodents (not forgetting Tiggywinkles,
Spineless and Spiny); – as croc or avian cochlea unwrinkles
At the sound of its hominid descendant; – as well-defined nervous systems
In annelids and arthropods grope the soil of neurological consistence
With human kinship; – as in vestigial prehensile tail we recognize
Our chameleon and sea-horse lookalike, with the same great rolling eyes.
But the song of climbing feet and tail has widest resonance
Among would-be ancestors of man, all claiming family resemblance:
Primates (monkeys are favourite), Edentates (sloth and pangolin),
Rodents (the arboreal porcupine), Carnivores (the palm-martin),
Marsupials (the possum). How discrete Orders could evolve same structures
(Like the gizzard in birds, earthworms, some fish) suggests an impetus
Subliminal enough to explain how incommunicado authors
Could go through identical stages of development of plot and characters.

But who dare provide a mechanism to account for *parallel convergence*?
Easier to ignore it, like Darwin, draw the tree of species divergence
From common ancestors, put man on the primate branch, to flatter
Most his intelligence and liberate reductionism for determinist patter.

Wisdom of self-knowledge is mocked by reductionists who look back to see
Whatever our animal antecedents suggested that we might be,
And is not repealed by the yen to make speculative generalizations,
Like Lorenz, Ardrey and Morris, from their animal observations
Of angry birds who swoop to strike off the crown from *Homo*,
Or of fighting Lorenzian fish who say only 'I told you so',
While imperatives of territory and survival float like an extended metaphor
On the current of Darwinian genetics, o'er the Freudian volcanic rumbler,
Making militant enthusiasm plausible, blood-letting the letting off steam.
I'm a holist, not a reductionist: I look first at what I became
And then look back to see what my ancestors might have been –
And confess I feel an affinity with the organically creative scene.
No mere projection of feeling is my deification of nature
But based on animal studies, of which Lorenz is undoubted master,
Who viewed the mythical 'beast' as a hypothesis to be investigated,
Not a dogma to be defended by the brute who exterminated.
Spare us no detail of criminalizing of animals by hunters' pride
And shame the viewers of *Jaws* by a guest appearance of Hyde –
But vice is not constitutional, but on the menu with good
And Jekyll had been a chooser, till he lost his taste for food.

The occasion of losing one's cool may ring like the echoes of rage
Down whispering caves of the heart, on which candour levies a wage
In guilt no behaviourist would pay (who holds all others responsible
For making conditions frustrating, all crimes being socially attributable):
A generous heart, conscious of how easily it can soil itself
With uncontrollable passions, will trade with its nature's pelf
To square accounts with non-cavers, who claim all aggression's reactive.
If the doc was aware of his natures, the good and the unattractive,
He could still rejoice in his discovery of the habit-forming functionary –
In powers of imagination and planning, by which he gave rein to Old Harry.
So what excuse can I have, if I find myself transformed
Suddenly by Hyde uninvited (in built-in reactions, unformed
By parents, teachers, subcultures)? While the beast has no foresight or tools

And worships the force of the present, salivating Pavlovian pools,
Man can predict and control his impulses on the biological level,
Transcend them on the psycho-social, where aims in means do revel,
Where all our doing has its fount, neither in the nature of the human
Nor outside, but in matters of conscience between myself and my fellow man
And between myself and the future I am . . .

6.

(With apologies to P. G. Wodehouse)

(The Scene: the steps of Berkeley Mansions, the country seat of the Woosters. Bertie and Jeeves are inspecting a large animal crate that has just been anonymously delivered.)

Jeeves: *Orang*, sir?

Bertie: Well, it certainly isn't Lord Ickenham's nephew!

Jeeves: Pongo? No, sir. However, by a curious coincidence
It is named *'Pongo pygmaeus'*; and I attribute the delivery
Of the all-too-human individual to Berkeley Mansions
To a pointed remark by big-hearted Uncle Freddie
That Pongo's the man to imbue a jackanapes
With the chastening humility of knowing his descent from apes.

Bertie: I prefer his impersonations to biological muck-raking. Rather!
And I'm darned if I'll believe that Pongo here is my grandfather!

Jeeves: Grand-uncle, sir, descending with you from his forerunner
In the Eocene, a prosimian stock that resembled a lemur.
But the gradual phasing of characters that the finches recite
Isn't helped by Darwin's demand for them to re-write
Gene codes at every bound, while mutations are charmingly
Illiterate in their disconnected effects and could botch alarmingly.

Bertie: I'm sorry, Jeeves, you've lost me . . .

(Jeeves takes a notebook from his pocket, writes the word 'Primate' and becomes engrossed in a word chain. After ten minutes Jeeves shows him the notebook, and he reads)

What is this?

'PRIMATE PROMOTE PROMPTS PROSITS
PROSIER CROZIER WOOZIER WOOSTER'
From Primate to Wooster! By jingo, Jeeves, that's clever!

Jeeves: Thank you, sir. It's a simple Darwinian statement:
Sense emerges, because species are functionally eloquent –
The evolutionary failures, the half-formed organs are conspicuously
Absent from seams – and by preserving the gist in harmony
With the broader movement of life I suggested Darwin's
Idea of the plasticity of traits. I'd have taken aeons
If I just rang changes – not airs, not Nature's carillons!

Bertie: But Jeeves was a metaphysical directing force to the puzzle,
Banishing blind chance, like Lamarck, Bergson, von Uexküll!

Jeeves: I permuted the words, as if Nature could read. So arise
Double-barrelled mutations, by virtue of which stepwise –
Though decimating the letters Nature uses – you 'evolved' from PRIMATE
In the Darwinian manner, gradual, progressive, seriate.
If the adaptive kernel is the interplay of countless genes,
Is the nut to be cracked by the mutant sledgehammer's means?
Such drastic measures clash with progressive aesthetics,
Too hit-or-miss, and falling short of genetics.
Can a theory of trends stand on mutant ground?

Bertie: Genetics is messy, as even Lord Emsworth found:
Black Berkshire sows cannot be made to measure!
All the more reason, when off the peg, to treasure
The occasional Empress – the suit that happens to fit.
Baggy or tight is the measure of Mendelian wit.

Jeeves: If the finer grades have eluded the breeder once,
They'll elude in a series. He applies the laws of inheritance
With such tasteless want of finesse – and yet he expects
Delicate trends to emerge when Nature selects
Blindly on the basis of mutant characters. Once,
For an Empress, would be miraculous! A prosimian dunce
Can't develop his wits by mutations unmastered in letters;
And the dearth of a theory of heredity adapted for trend-setters,
To explain the disparity in inherited variations and inheritors,
Makes 'survival value' a cloak for injustice. Did lemurs
Latch onto the importance of culture and intellectual improvement,
Never having thought before? Matched to their environment,

Why should they grow their brains? Ambition? A sorry
Reflection on lemurs is the thesis that their wits were too paltry
For the struggle and *needed* to evolve. Could *any* life at all
Be sustained by prosimians with a brain too crude for survival –
While a series of discrete mutations enlarged the skull
And, slowly but surely, the brain, the thumb, the mating call?

Bertie: Don't hold your breath, dear Pongo, as you wait for the dawn
Of your own humanity: the plans have not been drawn
For the mutuality of animal factors, the interplay of gene,
Of which we know so little, knew less in the Eocene!

Jeeves: 'Here are we, in a bright and breathing world:
Our origin, what matters it?' muses that inward eye.

Bertie: Jeeves, I'm much obliged for your felicitous reply,
Your third excursion into poetry! Tell me who penned it.

Jeeves: Wordsworth, nearly sixty years before man descended.
I took the liberty of quoting the bard, to prevent
A proneness to reductionism from usurping what is important –
To establish criteria for humanity, seeing through the slickness
Of the thesis of Darwinism, and to describe and defend our uniqueness.

Bertie: The original vision of man is rational, I guess so –
Now that the vision of original man seems less so.

Jeeves: Your definition of man, if I may widen it, dates
Back to Aristotle and kindred philosophical pates,
Whose Reason was activity, not something that qualifies activity,
Who outcast from *sapiens* the *homo* with a different proclivity.
The rational life, like that of the stoic or scientist,
Could be a placebo for the meaning and fulfilment he missed
When he failed to evolve the adaptive capacity in relationships
Or the gene for a passive vision of the truth, for worships
And the voyage of artistic discovery. These emotional and cultural
Abilities enrich life with discretion beyond the intellectual's
Sense of its pointlessness and tragedy. So I would tweak
The sapience in *Homo*, rightly thought to be unique
Among animals, to include all that we do in our very best moments
To make sense of life and the world with our natural bestowments.

Bertie: I can trust you, Jeeves, to do justice to the meaning of 'sapiens'.
Prizing our uniqueness enough, the animals we'll distance!

Jeeves: This begs the question, though: if the life of quality –
Attributed by the 'clever' to owning pots of money
And by the 'simple' to having a large brain – is a gift quite separate
From animal mentality, how did the human mind originate?

Bertie: I would venture to guess that the mind of man either mushroomed
Among the apelike grandchildren of Prosimia, or was primally groomed
For a fuller, more blessed life by the power of Venus,
The mother of creation, and so of the *Homo* genus.

Jeeves: Your Venusian theory of the origins of Reason, sir,
Is as plausible as anything an evolutionary biologist can offer,
Who infers from the threads, still intact in the moth-eaten strata,
A tapestry of evolutionary sequences and links without data.

Now a primitive fish, amphibian and reptile gird
For aeons of frustration, as they play their parts . . .
Landwards,
The *fish* must swim to grow legs, a crossopterygian,
('Not yet,' says *Ichthyostega*, the oldest amphibian,
'You've not even a pelvis!') and the supposed steps of metamorphosis
Would have littered the Devonian floor with peg-leg chrysalises,
Had the sea not been acid to transitional forms.
Swampwards,
The *amphibian* must climb towards reptilian ancestry, laying duds,
Until shell-suits, yolk, secretions to liquidize albumen
Could evolve one fine morning, with *allantois* for oxygen,
When the transition to terrestrial egg would be finally complete
And the emerging, reptilian sexuality of the parents could meet
At last with success, before their millionth birthday, as the first
Carboniferous reptile hatches, egged on by the cursed
Swamp – also acid to transitional forms.
Skywards,
The *reptile* must aspire to the feathered ancestry of the birds,
Fraying his scales into down, to warm up the Jurassic,
Hoping against hope to perfect the flight feather (the classic
Twinset of barbs and barbules that spreads out for lift
And clenches on the upbeat), the muscles and nerves to shift
Each aileron, the porous bones, the streamlining, the wing
From the forelimb, the one-way oxygen circulation for breathing
And metabolizing – with hardly a pant as it flies up in time

From the looming reptile, plopping in its eye the birdlime
From the air – so acid to transitional forms.
Bogwards,
The reptile must join the *mammalian* ancestry, turds
And urea excreting, far from uric acid dumps,
Straining to make the longitudinal muscle jump
Outside the circular in its gut and for its heaving chest
A diaphragm, an inspired companion to the emerging breast.
All this happened in the Triassic, the age that fed reptile
Hopes for mammalian offspring with mammae so juvenile
That the offspring cried for yolk, outgrown long since,
And starved, while the parents, a knotted and asphyxiated mince
Still working on a form of intra-uterine development, grew
Distinctly cold-blooded during aeons in limbo, to strew
The plains – which dissolved their transitional forms.
What cues
Must an evolutionist take, when the earliest, most reliable news
Of amphibians, reptiles, birds and mammals does not
Follow in the wake of the transitional forms that plot
The emergence of one Class from another? – when fossils tell him
That the Phyla, Classes and Orders erupted to quell him,
Radiant and blooming like Venus from the foam, clothed
In her finery of species and genera, with no part unclothed
Contriving to live and compete with organs in transition? –
When palaeontologists make the logical submission
That life-forms, unheralded, be boldly coloured like a Titan?

Taking as his cue any hint of homology to press
From fragmentary evidence proof of trends nonetheless,
He rides into the sunset on a form of progressive giantism,
The *Eohippus* to *Equus* sequence, where he ignores the schism
In the *Merychippus/Pliohippus* synod, when three toes united
Abruptly and grazers' teeth tried browsing mitres –
When rising numbers of ribs suffered reverses –
When therefore it was claimed the supreme ancestor precurses
The hyrax it resembles, a ride for cowboys discredited. –

So any descendant for a skin-deep lineage can be fitted

By interpreting the fossils to match the foregone conclusion
That an ancestor must exist.

Bertie: Will never his attribution
Of primate ancestry on fragmentary evidence collapse?

Jeeves: Hugging his own preconceptions, he will plead the gaps
And tomorrow's clincher. But the fossils themselves mock
That hope by the scarcity of 'transitional' forms and the stock
Of a quarter million terminal, the reverse of his wish: relative,
Not absolute, numbers stand as umpire and narrative.
Insofar as gradations are claimed for primate phylogenies,
Certainty lies only in attempted disproof of hypotheses.

Bertie: You'll pop them one on the jaw with that one, Jeeves!

Jeeves: Few will explain why primates should spring from a prosimian;
Why *Aegyptopithecus,* the swamp-ape, should be allowed an opinion
About being ancestral to man when drier, more forward
Primates would not have a pickle of a chance to be heard;
Why his son, a *Proconsul* in the Miocene, who was certainly no swinger
(Too stiff in the wrist), should still be our mainstream 'ancestor'.
His own son, *Ramapithecus*, was into mysticism
And the Pliocene, wore round his molars an unmanly cingulum;
A masticator, they say, uninstructed in the use of a pebble-chopper
On meat, he acted and looked like an ape. Of his whopper
Brothers, one was pongid, *Sivapithecus*,
Another was a bamboo-eater, *Gigantopithecus*.
Some say that tool-use may later have made apes petite,
But no proof was there yet that any ever walked on two feet –
Till Lucy from afar gave primacy to bipedalism – and tree climbing
(As curved phalanges prove!): unwaisted, pot-bellied, chiming
With funnel-chested, brachiating but toddler-stumping, hardly a prototype
For arm-swinging, hip-swivelling man! But Lucy was of the type
To be humanized by bipedalism alone, the artist's peccavi. –
A robust form might later eat plates, a gracile sip gravy:
Could a light-skulled, culture-bearing Southern Ape be the stepping-stone
To our striding gait? for we can hardly suppose a bone-
Head *africanus*, with chimp brains squeezed by an eye set high
In a skull, jaw-slung, to see boughs, could learn to defy
Its hams by footing it, upright. His stiff-shoulders connect
With a gorilla-like trunk, craning his long muzzle, bedecked

For blinding – then, pin-buttocked, he drops down for four-square vision
To knuckle-walk the plains, sun on his back. But the decision
Of the Pliocene Southern Apes to replay the card '*robustus*'
At the dawn of the Ethiopian Pleistocene with machismo, just as
We come up trumps and the grace-card proves to be the Queen,
Reads like the homecoming of a drunken beer-belly, the scene
Where the woodenness of the walk betrays the stiffening of the brain.
This doomed intermediate between man and ape, it was plain
From multi-variate analysis by Oxnard of Chicago,
Was uniquely different from either, and exposes the farrago
Of theories of descent, ossifying in comparative anatomy.

Bertie: Is the field now clear of pretenders to human ancestry?

Jeeves: The field, not the circus – where in the quest for origins
The molecular voices of the past are gathered in
For a spectacular tossing of options for a family tree,
Sprouting the closest members of the hominoid trichotomy,
The chimp, the gorilla and man. The performers' questions
Are small, but they make their answers large, their suggestions
Flung up like jugglers' clubs through the windy air.
Twin fliers, in the metaphor conceived for the likeness they share,
Are aloft, while a third is poised for a different trajectory.
DNA sequencers schooled their clubs in the ambiguity
Of letting the gorilla and chimp ride the air together,
Separate from their human brother, and from the same data
Letting the chimp and man ride, while the gorilla is apart;
Such sleight-of-hand of analytical procedure could start
A battle for primacy, so other molecular data
Were used to isolate the chimp and twin man and gorilla.
Some jugglers put the chimp with man on the path to glory,
Demoting the gorilla, without moving from their laboratory,
Using the methods of DNA-DNA hybridizing,
G-banding analysis of chromosomes and protein sequencing,
While others, as adept, made blood-brothers of chimp and gorilla,
With phenetics (loosely sorting the morphologically similar),
Restriction enzyme analysis of chromosomes, fossil,
Molecular and morphometric analysis and studies biomechanical,
Cold-shouldering man. (C,G)M was the likeliest
Throw of the three – but (C,M)G was the cuddliest.

Bertie: I always said you were ahead of your time, Jeeves –
But this is ridiculous!

Jeeves: I'm sorry if my erudition grieves
You, sir; but if it could save you from a Darwinian position
That neither of us find palatable, to wit, the submission
That man and apes were descended from a common ancestor,
And if I could inspire your confidence in my conclusions, sir –
Would you thank me?

Bertie: I would, Jeeves! Speak, I'm all ears!

Jeeves: Dr Alexander ('Chimp') Twist is as slippery as smears
Of axle-grease could make him, as you know. Now he schemes
To grab over 98 per cent of human genes:
He closes his hand, like the clasp of chimp DNA
Around human, 2 per cent mismatched (2 degrees away
From the annealing temperature of a mirror image) and steals
A fistful of *identical, aligned characters*. So he feels
It vindicates the homology thesis, invoking his ancestor!
DNA is a twisted zip, a chromosomal jester
That grins multimillion teeth, christened 'base pairs'.
But the pride of ancestry Chimp stole can give him no airs,
For much gene profile is padding, while the 2 per cent
Of human genome, that bust his zip, had bent
Two million of his teeth, so loud with signals of distinction!
Some say that our distinction is due to homologous regions
Losing identity with the time-lapse since Speciation Day
When we said goodbye to Gramp Twist. But law's the mainstay
Of realist science! The discovery that congruities exist
Between genomes should not have invoked the story of Twist
But hinted at constraints on possible genomes imposed
By *general principles of organization!* It's also been proposed
That to explain the facts of homology by history and history
By homology is *circular reasoning*, assuming an ancestry
In order to prove it. Now the proper approach to the correspondences
Noted between man and chimp in their DNA sequences
Is to refer to the compound eyes of crustacea and insects
And other similarities in organisms no evolutionist connects
By common descent (like hawkmoths and hummingbirds, the *tracheae*
Of insects and spiders, the animals and plants having *ciliae*,

The flying phalangers and squirrels, the frogs, lizards
And tree-snakes that glide, the birds and fish with gizzards)
And to explain the congruities in terms of the adaptive success
Of universal laws of form and morphogenesis,
To account for what cannot be inferred from an ancestral condition.
The penchant of complex systems for self-organization
Is one that each knows with his mind and heart, and premises
A universal adaptational programme on an 'analogue' hypothesis
Abler to explain the independent acquisition of like organs
And functions in lineages as distinct as insects and crustaceans.
The *macro*-evolutionist compares organs for a verdict that bores
The *micro*-aligners of DNA sequences, who cause
Their own 'homologies' by begging the question of descent.
The bat and the dolphin both have sonar, meant
For themselves: as neither has a hand-me-down from the other,
Their lineages are *parallel* to Mac. Yet the human character
Is *derivative* to Mick. The bat and dolphin were there
When the parallel branches were handed out, but where
Were we? In Mick's own little world, on a three-branch tree,
Where congruence between data-sets can never be reached independently
But receives a common explanation, for the sake of 'parsimony',
And an apt 'weighting' of characters to give a Twist to our phylogeny!

Bertie: That's rum! If they can't distinguish between homology and analogy
Objectively, we could all be barking up the wrong tree!

Jeeves: 'Alike' and 'akin', confounded, will assemble a brother
From a scarecrow to endear the unknown horizon and smother
All differences, till we wonder how we came to lose our straw
And must-glands, and find our feet – and find the flaw,
The habit of leaning on brooms, the vacant stare,
As pointers to our origins!
Just so did the maiden fair
Bathe away the fur and scent glands from her ape-like body
And adapt the eccrines that kept her from slipping on the wadi
For overall sweat-cooling, till the sea-ape had human skin.
Then heat-stroke and dehydration are adduced as the reason
For calling our sweating an unfinished and wasteful experiment
By the self-depilating beached-ape – which forgets the sentiment
That our acclimatization to heat extends to behaviour
Like resting in shade and washing the pores of the saviour

(For liability to blockage, not secretions, too few not too many
And ignoring the signals are the cause of heat stroke), while a penny
Spent for water-loss strikes no one as life-endangering.
Our heat-regulation is *perfect* and *unique*, majoring
In histology – only flunking in the hot and hazy view
Of some fur-to-skin theorist!
Just so, too,
Did the bathing belle, barish like a seal and sweaty,
Grow sebum-watertight and warmly, grossly fatty.
Then ten times more fat cells than expected are adduced as a reason
For saying that our skin's a vestigial sealskin – a treason
To the lean savannah-ape theory and slimmers' pride!
As this twaddle is mooted, lipase breaks down triglyceride
(Turned on by noradrenaline) releasing from the cell
Fatty acid and glycerol the more we exercise or thrill.
Is anything wanting for the conversion of triglyceride in this landlubber?
Human obesity is due not to inherited blubber
From an anthropoid seal, but to the torpor of the manatees!

Just so, as Ethiopia flooded, we reared up with ease,
Hands free to manipulate our watery environment, bending
At hip and knee like a chimp walking (expending
Little energy while hands are held), waiting for balance
And hip-swivel to develop, to allow us to stride the expanse.
Bad backs and varicose veins are adduced as the reason
Why bipeds evolved in the sea, for in the dry season
Aussie Jellylegs only *shambled* (its bipedal mood
Was the chimp's half-crouch, hamstrung if ever it stood
Like man, unable to bring its leg back), a pro
At waddling, but disabled for the long slog on land – which the physio
Reminds us *cannot* be our element, else our pins are not dodgy!
Absurd! More distinctive than tool-using or dental morphology,
Our *perfection* and *uniqueness* is human locomotion, *savannah*-style!
The momentum of the forward-swinging leg moves the torso a mile,
Knee-locking, hamstring planting the heel, gluteal
Controlling the vault. Swinging a leg is ideal
For taking the thrust off the glutes! Such an energy-saving
For a creature walking on land is lost by wading!
Bad backs aren't relics of an aquatic capability, but a plan

To disdain the terrestrial, the easy stride of man!

Just so did the wrynose obtain a wrinkled skin,
Then searched for the pattern her own preoccupations predestine.

Bertie: For sure, you keep science on a short leash, Jeeves, with your knock-down
Arguments! But I know of a chap who loves Pekes, who would frown
At the claims you make for humanity's unique progression.

Jeeves: It would be boastful indeed to pride ourselves on the possession
Of qualities in which animals outshine us: cheetahs are more fleet,
Beavers are more diligent, dogs are more loyal and discreet.
But surely it's wiser to obtain the measure of mankind
From standards uniquely applicable to him, than to find
Animal criteria to apply, reductionistically?
And surely it is humbler than talking of origins, sophistically?
Now what I'm obliged to contemplate, sir, is the provision –
Or rather the lack of it – that science makes for the cognition
That the proper study of mankind must surely be man.

Bertie: Right ho, Jeeves!

Jeeves: Presume not genes to scan,
Reducing the archetype to a beanbag and the claim that egotism,
Spite and sexual stereotypes are driven by them –
And then plead immunity from questions of human rights,
Human dignity and the sacredness of life and its delights!
Our uniquely human gifts have much to do
With how we perceive and treat our fellow. As Bijou
Has olfactory gifts that shape his social awareness,
Our uniqueness is a socializing matter: it lends a fairness,
A moral and cultural value to the semblance of humanity
That demands to be judged by individual contribution to society
And human welfare. Or are we a race of base pairs?

Bertie: The spawn of dragon's teeth, laying in the dust our airs?!!

Jeeves: The human genome is not the genome for humanity:
Soluble in mankind's gene pool is any kind of integrity
To unravel the cultural strait-jacket DNA weaves –
To hide and restrict humanness to what investment achieves
In jigsaw design.

Bertie: Once lost, the photo on the box –
And it's guessing by colours and thumping the makeshift blocks!

Jeeves: 'From nature's chain, whatever link you strike,
Tenth or ten thousandth, breaks the chain alike . . .
All are but parts of one stupendous whole,
Whose body nature is, and God the soul.'

Bertie: To whom do I owe this stream of verbal felicity?

Jeeves: To Pope, sir. It's funny how the hope of 1733
Could become the light of your butler who fears the satyr
Of planned genetic change two centuries later,
Who scarcely guessed, after reading *The Genetical Theory*,
That four lines in An Essay on Man could frame a query
Outliving genetics and in rhythmic measure illumine
The highest that the heart attains with the uniquely human.

Bertie: A *Homo* comes to sapience and flunks the apes!

Jeeves: He comes to a unique self-awareness! Though Darwin called apes
Our 'brethren in pain', bequeathers not only of our pedigree
But of mental experiences we are bound to infer from animality,
Of pain, hunger and rage; though the ability to detect
Food or enemies and signal it Nature would select –
No advantage accrues to knowing that one is aware
Or guessing what others may think, to being self-aware.
Can we attribute self-awareness to Pongo with equanimity,
So forgetting ourselves, gifted with that unique susceptibility
To guilt for evils (like habitat destruction) we foresee
But do nothing to prevent? Inasmuch as we disown it in ourselves,
It's no remedy to anthropomorphize apes or machines! As man delves
In the cellar of his outcasts, he meets his brother there,
A fellow-creature in danger – and himself aware,
And he feels compassion. Is this an intellectual inference
From the affections, hopes and ideals of *Homo sapiens*?
Or an instinct, inherited from apes, for social communication?
Do high moral values have a symbolism and mode of action
On a scale with primate grooming? From observations it's evident,
In the absence of introspective reports, that a conscious intent
To speak on cue about the ripeness of forest fruit
Marks our correspondent in Borneo. – But postings won't suit
To far-distant places and times; he will neither prevaricate

Nor coin new messages for comment; his gestures won't integrate
Untrained into sentences, the means by which cultures evolve
From the nuanced signals on which learning and reflection devolve.

Bertie: But our train to Athens is still in the shunting yard,
The signals are garish and the apes are grunting hard!

Jeeves: Apes only in name. When the reductionist trance has been broken,
The power of our culture to keep us arboreal will have spoken.

Bertie: I see! For 'apes', read 'barbarians', Jeeves!

Jeeves: As a good
Athenian, I enjoyed this dialectic and trusted you would.
But one fact ignored could render my whole case unsound:
Pongo's relations are not so thick on the ground.
Self-aware, we must do whatever is best for him.

Bertie: By becoming his 'brethren in pain'.

Jeeves: Let that be our hymn!

II

Adam's Grim Progress

'With regard to early hominids, Australopithecus was in the ape pattern and H. habilis (KNM-ER 1470) the human . . . It seems that relatively small changes accompanied the emergence of the earliest available and analyzable hominids, the australopithecines. These comprised a minimal increase in absolute and relative size and some limited reorganisation of the overall anatomical structure of the brain. As for neurologically important changes in the brain, there is scarcely any evidence of surface alterations in the sulcal and gyral patterns . . . However, major expansion of the brain and critical cortical reorganisation were striking features of the postulated transition from A. africanus (or an A. africanus-like form) to H. habilis. These changes included notable augmentation of the cerebrum, strong lateral expansion of the parieto-occipital region, the appearance of a human-like sulcal pattern, and the emergence for the first time of protuberances overlying what are interpreted as the anterior and posterior speech cortices . . . The superior parietal lobule is well developed and . . . may provide evidence of a functional asymmetry in the representation of visuospatial discrimination and judgment . . .

'Modern humans form a highly variable species. Homo erectus is similarly marked by appreciable regional variability, and I believe such polytypy is there at the level of Homo habilis. It is a feature that was made possible by man's culture. Our speciation and evolution are different from most other species by virtue of the cultural dimension. This has enabled us to diversify all over the world without speciating, unlike other creatures.'

Tobias, P. V., The Brain of the First Hominids. In J. -P. Changeux & J. Chevaillon (eds) Origins of the Human Brain, (Clarendon Press, Oxford), 1995, pp. 60–61, 83.

'Yahweh God said, 'It is not good that the man should be alone. I will make him a helpmate.' So from the soil Yahweh God fashioned all the wild beasts and all the birds of heaven. These he brought to the man to see what he would call them; each one was to bear the name the man would give it. The man gave names to all the cattle, all the birds of heaven and all the wild beasts. But no helpmate suitable for man was found for him.'

Genesis: 2:18–20 (Jerusalem Bible) (Darton, Longman and Todd, 1968)

God gave him to the earth to serve primal forces,
Habile, but unable in naming his pets
To live up to the ideals of the Golden Age myth
That earmarked him for failure two million years later
And for investiture with an ape's inflationary brain.
But with a workshop in his head and words in his heart
Adam could have designed him or called him like Tarzan,
Naming him 'Cheetah', that australopithecine!

Nature was the tyrant of his fate. Small wonder
She let him think that the ground was accursed,
Where God might have walked in a swirl of leaves,
The moment he asked why rights and duties,
Blown in, should root his moral judgments,
When the Tree of Life grew up to bring
Far dreamed-of altruism from its cultured soil
To human sight and sanctify life.

He knew too much, asked more than Nature
Had the power to grant in the kindlier mysteries
Or his tools to remit from the struggle for life
And left Eden for a cruel, disillusioned future.
But I defend his right, invoked by the myth,
To fix priorities like a god, preoccupied
With the goals of life, not the trappings, to be ruled
Less by expediency than intrinsic value.

Naming the animals Fragile and Resplendent,
Modern Man has arrived, not pushed from behind
But drawn by the invisible bands of culture.
Growing in silence, nearly come of age,
His restless senses declare a truce
With Nature: having mastered his warring impulses,
Brought desire to parley with reason and spirit,
He no longer wants his choices made for him.

Such equanimity is the Tree of Life
And freedom of spirit, post-Edenic Eden.
Many thinkers and artists have known it,
Ascribing to leisure, comfort and money
The status of means to ends, not ends
(Whereby we are enslaved by nature, not freed).
Their disinterest annotated the steps others took
To stay Nature's pinch and prolong play-time.

One and a half million years ago
Adam strode erect and contrived flake tools
To replace his pebbles. Though his brain had deepened
To industrial lordship, still no quarter
Did Nature give to idealistic wimps.
The ends we brand utilitarian, pivotal
To survival, like safety and might, preserved
His line, until *Homo sapiens* could choose.

From the hive of culture not thickness of skins
But quality of thoughts we take. This honey
Was for tea, not breakfast, to sweeten our lot,
Having lost all innocence, as we make our choices
Between instinct and altruism, expediency and morals,
Animality and self-awareness. Bare Adam took the pragmatist's
Part and thought with flint, that sparked
A culture, destined for more fastidious tastes.

Creation grew conscious in Man, lately
Rocketing, culture-fuelled, from the dominion of Nature
To an orbit beyond the self-wrapt and material
Ego. After aeons of wresting from Nature
A meagre existence by the sweat of low brows,
Are we now embracing our common destiny,
Or is the voice from the ends of the earth
Still saying, 'These flints have a fine cutting edge'?

III

Homo Frustratus

(For Hojeong Kang)

The computer cups its hands around nothing, as peaks
Of microbial bingeing flee from their anticipated measurement:
The goalposts have moved yet again, as my new column wreaks
Chromato-havoc. I am Time's conquest. I waive my own nourishment
To catch up. A suspected short-circuit sets my biker brain
Racing with the prophecy that either I home without lights,
Too late for buses, too skint for taxis, or I gain
By footing it six miles for the poetry my blood indites.

My mind opts for movement, my body for the opposite,
For stultifying avoidance, as it struggles to grudging admiration
Of *Homo erectus* – but not as libertines would have it –
For his capacity, simply, to stand up! My recent saltation
Up Snowdon's rocks, for my health, did severely tax
My vertebrae's knowledge of their comings and goings, and I'm dogged
By the memory of my sturdy comment to Bertie that backs
Are biddable on land and needn't have evolved waterlogged.

In pain, I appreciate the sea-ape, and doubt the benefits
To Cro Magnon Man of an inspiring, though moonless, perambulation
Marred by Neanderthal footfalls. My battle of wits
With gremlins (with its implied, mythic re-orientation
To motorbike electrics and that other persecutory instrument)
Proves man self-aware, warring between hope and fear
In imagination, the better to adapt to his environment. This intent
Makes Man the lord of his domain, *il va sans dire*.

That manifest pinnacle of human achievement, Science
Is for meeting difficulties constructively, with fortitude, viewing
Pig's-head exorcisms in Korea with rational defiance:
Mere fancy is my notion that a hubristic computer is imbuing
Equipment with a saboteur's will . . . I'll type Adam's poem,
While the wheels of research are turning, sceptically paced.
So I creep to the computer for distraction and to tap from the phloem
Of the Tree of Knowledge the reasons why we are graced.

Now is the voice of human adaptability drowned
By a thunderclap, as power failure strikes, downs tools in sympathy
With a headlamp and wastes an evening! Things dumbfound
The ability to judge and solve problems, if compromised by penury,
Pain and stress. Human poise gone, leisure
Is not means, but unseemly end. It is light. I ride home
And stare at fantasy and paranoia. Still, there is pleasure
In mature self-awareness, though its cost is frustration to some.

IV

Echoes of Science Opening its Head

Dedicated to Augusto and Michaela Odone and the makers of the film 'Lorenzo's Oil'.

(For David Dowrick)

Science: Don't come into my laboratory, Maude
Or let Junior tug at my coat!
Who knows if he will be awed
By this bird I hold by the throat
In this test, with a scientific ring,
To see how he spreads his wing?

Echo: To see how he spreads his wing,
This man who's reached his true self,
Recalls the all-powerful king
Who leaves his people on the shelf,
To starve on guesses of his visions
Or to blindfold them with divisions.

Science: The thinking behind pure research
Is more open than the mystic kneeling.
Yet I admit, in the quantum church,
I achieve suspension of the feeling
In common with the mystic race –
The feeling of time and space.

Echo: The feeling of time and space,
Freed from mean expectations,
May repair the veil of grace:
To the opened eyes of the nations
Man's is more vain than the sublime
Signature in the sands of time.

Science: But it's nature's purpose opens
Full-blossoming awe in my heart –
Though incubating that sense depends
On my pattern-discerning art
Rooting in the dark journal's soil,
Heated to nearly blood-boil.

Echo: Heated to nearly blood-boil
Is the hell of high anger and fear,
When ignorance and ambiguity oil
The wheels of delusion in the seer
Of patterns: the witchfinder's leading
Was *scientific* in its special pleading.

Science: But I'm wary of his dangerous certainty
And the truth behind semblance I entreat
With caution and objective theory
And with all that tests can mete. –
Yet the quest for harmony is votive,
And so any debate is emotive.

Echo: And so any debate is emotive,
When the delusions that science indicts
Have the heart's confirmation, denotive
Of the popularity for which scientist fights.
Yet interest in the occult may be creative
Action for a purpose that's sanative.

Science: People can be sane without delusions –
And I ought to give dreamers wide berth;
I do get so angry with the effusions
Of people who reject the worth
Of scientific findings. I suppose
The intellectual hatred shows.

Echo: The intellectual hatred shows
In Ulster, witch-hunts McCarthyite
And Rushdie's fatwa: it grows
From the compulsive need to be right
And consistent, as we leap to conclusions
We incline to reach by exclusions.

Science: I admit that I try to make
A thesis from inadequate data –
But importantly for elaboration's sake
Referees give their imprimatur
In order a descant to permit
On concurrent facts that confirm it.

Echo: On concurrent facts that confirm it
The paranoid will ceaselessly batten,
Till his fear is 'proven', as a hermit
Spins his world-tottering pattern
From the exaggerated evils he'd impute –
And *you* pattern your world, to boot!

Science: I'm open to *dis*confirmatory evidence
As well, being involved in the world. –
It's just that my thesis has a defence
At which no charge can be hurled:
My opponent's methods were duff
Or he didn't understand me enough.

Echo: 'He didn't understand me enough
Or he was lying to me';
'He wasn't born from above
And I have had visions, you see';
'You're in the power of Satan,
Whose delusions severely straiten.'

Science: I walk in possession of the sun
That thaws the glaciers of theory,
The fleeting rays of speculation
Upon every factual query:
If those beams had no mandate to endow,
I would abandon my theory right now!

Echo: 'I would abandon my theory right now,
But it's a fancied high-flier to others,
As it leaps from the evidential bough
To the wing-beat plaudits of my brothers,
Happy, *clappy* to the rainbow,
As high as orthodoxy will go!'

Science: I don't stand aloof, like religion,
From upset in the fray of discovery!
By elevating the spiritual like a pigeon
To carefully nurtured mystery
Launching from Serpentine's edge,
It shows it doesn't value knowledge!

Echo: It shows it doesn't value knowledge,
Reducing soul to brain's wiring
And blunting the ethical edge
That stops your mind from enquiring
Why you're prey to indoctrination,
The fallacious reasoning of Reduction.

Science: The deep answers of sages and religions,
Like the soul, want for the tangible
Universality of science: the belief-systems
They spawned are de-constructible
To leave room for varying interpretations
In the hearts and minds of the nations.

Echo: In the hearts and minds of the nations
Doubt has its place: the thought
Is all the more ironic that affectations
Of utter conviction are caught
By the young from scientific teachers
With the missionary zeal of preachers.

Science: I have no doubts at all
That science will heal our tomorrow:
It perpetuates no tribal brawl,
No deadening tradition of sorrow;
It's the answer to disease and poverty.
This is my dream, sincerely.

Echo: This is my dream, sincerely:
A just distribution of privilege
And teachers – not scientists merely –
Liberty from the tyrant's bondage –
A more practical and sympathetic view –
Are all part of the solution, too.

Science: But scientific solutions are cross-cultural
And speak with universal authority,
An all-explaining cure-all
For the ills of ill-informed humanity:
Scientists are true heirs of the Renaissance –
Nothing shall hinder their advance!

Echo: Nothing shall hinder their advance,
When Milgram's saints march in
On the experiment to elucidate the response
Of unquestioning obedience to a boffin
That manifests in administering shocks,
Proving deification of docs.

Science: But nothing must stop the patient
Gathering of corroborative data!
Nothing must stop the patient
Dying from the diet we administer,
Till statistics *show* it has failed
And more homespun methods can be unveiled!

Echo: And more homespun methods can be unveiled,
Such as in the dietary case
When a father researched what availed
In rapeseed oil to pace
His son's fatty acid production
To his myelin sheath's construction.

Science: They thwarted the experimental run,
The Odones without a PhD
Who wanted to save their son
From Adrenoleukodystrophy –
But the bringing of data to fruition
Is our collective mission!

Echo: Is our collective mission
To foster scientific hacks
Or interdisciplinary speculation
About the value of facts?
The thought is very preoccupying
That my sense of priorities is dying.

Science: It's not hope that gives value to data
Or love that remyelinates lives.
There's now no treasure greater
Than the comb of empirical hives,
Where the competitive workers thrill
To the merry jingle of the till.

Echo: To the merry jingle of the till
Pawns dance for protocol's sake –
As commercial decisions kill
Objectivity for objectivity's sake.
Numbness and paranoia I scan
In the pattern-seeing potential of man.

Science: In the pattern-seeing potential of man
Is the breeding of suspicion in *you*:
Religion is the divisive madman
That claims a total view.
While science claims oh-so-cautious progress,
Yours is the grandiose guess.

Echo: *Yours* is the grandiose guess!
You distance yourself like a mystic,
While you seek the nations to bless.
Your thoughts are those of a fanatic:
You make large claims, yet protest
Your respect for the factual interest.

Science: Your respect for the factual interest
And evidence has never been strong!
Your thoughts have obsessed and oppressed!

Echo: *Your* thoughts have obsessed and oppressed!
You are oppressed and obsessed!

Science: *You* are oppressed and obsessed!

Echo: *You* are oppressed and obsessed!

V

Manscape

Fine prospects step out of the past and perplex
The artistic yet rational eye with demands
To pay history-book homage to the springtide wrecks
Of Llewelyn's hopes at Edward's hands
Which bloodied the hills where the tourist stands.

The walkers' thin superiority assigned
A destiny to mountains unforeseen in lore –
That the oppression of man be out of mind,
Out of sight with map and flora,
Bleating a privileged pastoral of awe.

My spirits are oppressed by the weight of the world
As we drive, bloodless and science-enchanted:
Gwynedd and Powys have their flags unfurled
To welcome us spectres at a bug-feast, decanted
At Plynlimon to sample the gases they panted.

The hills have no voice to arraign the English
Or tell how sheep farmers have fared and sweated;
But I know that they hoard bright memories and anguish
Woven from expediency and compromise, and have netted
For memory's pocket a wealth unregretted –

While *I* must dig deep for a penny of value.
Leisure need not be wayward to miss
The symbolism of landscape, the dragon in you
That shapes your lair to yourself and the bliss
Of repose in four corners of legend and artifice.

Beauty is found in the banal and habitual:
Visiting Dutch, though intrigued, aren't enamoured
Of the lack of flatness they grew up with; casual
You don the tracts, threadbare of sward,
Seamed with water – not the fashion abroad.

If mannerly clothes makyth man, if true
And homely surroundings are a close-fitting hood
That we wear to preserve the self we value –
The town and country planners only could
Give a logical account of the aesthetics of mood.

What they have shaped ranges from the unlivable
To those humane schemes that are lucidity's friend:
Composition needs discernment to select what is valuable
And the crystal words that through woods did wend
Down Wordsworth's life were no symmetrical blend.

The reminiscent way of beauty and innocence
Makes glimpses of home, after miles of harshness,
A bath for the eyes, a mystic suspense
Between real and dream, when we feel the caress
Of sanctified vision and the long hours we bless. –

And what calls me to Plynlimon? Not instinct, but analysis!
The Reason, that sifts our minds like wheat
And tosses our values and beliefs into emptiness,
Has not the skill to beautify one street
Or from things of worth to construct the sweet.

My stare, so shameless, the mountains return
And answer, like a mirror, my bullets behind glass:
I'm brain-shot with disharmony and harmony, in turn,
As weed-spotting eyes make a tourists' farce
Of the kitschy wallpaper of the Beddgelert pass –

None stranger than *Rhodo*, or homelier to stranger!
I thought I could paint a better canvas with my eyes.
So, all journey long, they become the arranger
Of aesthetically compelling features, like skies
Flecked with cotton, cypresses that disguise

Hard-edged buildings, tree borders to the road
Giving glimpses of reservoir, the patchwork of conifers
Glaucous and discordant against the ochre and sheep-mowed
Uniformity of moor, the textures of furze,
The profusion of geometric shapes and colours,

The exposed bleak summits, the windmill's personage,
Forested ridges stinting the moor –
Then the slough that completes my aesthetic pilgrimage
And embogs my tentative efforts to explore
A self-renewing beauty that enters by the back door. –

And all that splendour just ended in valuations
And abandoned brushes! Love alone paints loveliness,
Colouring in its theme from the palette of associations
With the paradise of memory or taught by neighbourliness –
And love goes West without the pain of homesickness.

Doubtless, self-exile would quicken the visionary
Gleam of Eden and some Celtic fervour!
But added to the interplay of perception and memory
That adorns our now and encourages us further
To count each blessing, is a hope preserver:

This is realization that our art subject, like life,
Is full of neat surprises that shift
Our vista of the present and save us from the strife
Of creation; for the patterns we see uplift
Never so well as Nature's gift.

Health and curiosity now sink to their eyes
In a *Sphagnum* bog in the quest for harmony
With the mystery of Nature. Closer than surmise
About the earth I touch, earth touches me
Suddenly, where I linger in gloom and lethargy.

Symbolic, a chestnut grace becomes
My eloquence: 'Kite! Over there, a Red Kite!'
Milvus milvus plucks me from doldrums,
Proclaiming its come-back in a victory flight
From near extinction from British sight.

Just hanging on by its dinky talons –
Pairing for life – eyes eight times keener –
It certifies all beauty, and has the talent
To make my pigment eight times greener
For re-enchanting the world than a critic's demeanour.

VI

The Shape of Nurturing to Come?

(To Angelika Steininger)

A baby, bawling for refreshment, warmth and dryness,
Turned sociable at the sight of instinct's smiley face –
Hovering – complex – full of interest – nighness –
And cooings – rationing his understanding of outer space,
Opening channels of thought-transfer between adult and infant.
His mind, specialized for seeing patterns, began
To shape the thronging impressions into a coherent
Picture of caring behaviour, a *gestalt* of Man.

And his *kindchenschema* stimulates our nurturing response:
A rounded forehead and body, dumpy extremities,
Soft and podgy shell, plate-eyes, big bonce,
His *gestalt* is far more than an expression of cuteness, it is
An icon of plasticity, cerebrally etched, and the capacity
For growth. Not inherently under instinct's or distrust's reign,
His needs for self-awareness and fulfilment are part served by stimuli
Of a two-million-year-old cognitive bulge in our brain.

Were faces a craving, I'd not have been bogged down in the *gestalt*
Of landscape, oozing from the habiline's cortex; but your word,
Angelika, was Brünnhilde's, charming Wotan by default
Of recalling his inner femininity. My anima was stirred
When you asked how eye-contact with a child can carry a value
To mobilize a father's tenderness, his anger to disarm,
While current affairs can barely kindle this value
In an eye, glazed over by the more obvious power to harm.

I framed your questions at the ice-cold bar of reason,
And Lorenz, your countryman, answered the instinctual part
(In the arms of your concern, his ethology humanizes in season).
For the glassy documentary eye, I gathered my *gestalt* heart
Up in my fist to thump out on a rostrum a picture
Of social reality that we all simplistically share:
To grasp the complex, we draw bounds between neighbour and stranger
With polarizing distinctions, like 'I/other', 'we/they', 'here/there'.

We draw the lines for an architectural plan in the mind
Of rooms within rooms, sealed with categories of our embracing,
And serene we sit in certainties of the inmost kind,
Either aching for more global empathy or straining through the biasing
Filters of prejudice dignity and the bonds of humanity.
The inmost walls of a bankrupt Austrian farmstead
Rang with the annexer's praise, when he tore up the chitty,
For he was German, not Russian, and their truth was bread.

When Adams and diddums were budding, all that was pukka
Were living shapes, patterns of bounty and categories
To hold the beautiful, remembered times of succour,
To round their *gestalt* before they slept – in paradise.
Science and poetry were bedfellows. When infant eyes
Met father's, they laid bare his emotion. But as he grew,
Another's mind-slots held him behind screens of ice:
'Minor', 'street kid', 'Third World' – no longer peekaboo

But a categorical outsider, as state-bureaucratic distinctions
Are used to justify neglect, or edged to express
Prejudice (or the predilection for false or limited information).
Truants, sweatshop drudges and war orphans bless
Each mission into outer space. The gulf of silence
Breaks with the piped intelligence of a cognitive fault:
It speaks, not of politics or mind-slots or poles of convenience,
But of unique individuals and events, no part of our *gestalt*.

The innocent escapee from the Procrustes' bed of our categories,
Who resists labelling in terms of our conservative *gestalt*
And cannot be stretched or lopped to suit, hurries
From the stratosphere to give the glad eye, and to tax and exalt
Our ancient, corrective centres of speech and simile:
'They' are like 'us', 'we' could be 'they', 'alienage'
Could be 'fosterage', the pairs switching roles for our *gestalt's* remedy –
A dichotomy added, 'man-made/made in God's image'.

VII

Homo Sociologicus (subspecies, *rationalis*)

'Many have original minds who do not think it – they are led away by Custom.'

– John Keats

'A society made up of individuals who were all capable of rational thought would probably be unendurable. The pressure of ideas would simply drive it frantic.'

– H. L. Mencken

Ours is the infant nature
That cannot turn away
From the smile that suckles us,
But makes our refuge –
And our adult shield –
The one who helps us interact
With an unfriendly world.
Keats's mind was social
Before it was individual.
But Fancy and Poesy
Aren't what the masses
Think they are.
Keatsian Nietzsche would turn
The Age of Ignorance
Into the Wisdom of Ages
Unravelled from great skulls.
But these were buried in the potter's field,
The biblical Field of Blood,
The end of egoistic aspiring,
For the potter is the would-be
Shaper of destinies, the Utopian
Like Rousseau, Kant, Goethe, Marx.
Darwinism! that expedient to place socialism
On a rational and scientific footing
By making history progress
By class struggle to the classless society,

Its post-industrial end!
The end of history
Came sooner than expected –
From the disabuse of an unripe proletariat
Undergoing rites of passage
Meant for the years of discretion.
If man were a rational being first,
Then social,
If his social qualities
Were improved by his use of reason –
Ceramic utopianism,
Individualist or collectivist,
Might have produced
A Grecian Urn by now
Or bricks.

Keats, Nietzsche and Marx contrasted Custom
(The arbitrary and capricious fashions,
The enervating norms, the unjust institutions)
With Nature
(The universal capacity for acts of red-blooded
Rational assent and dissent)
And sided with Nature against Custom.
Mencken, for the Moulded-by-Society School,
With officers of the ilk of Tönnies, Durkheim and Yeats
(Who prayed for his daughter), manned Fort Custom
Against Nature.
My mind was for Keats *et al.*,
My heart was for Mencken.
Not surprisingly, when the outlaw rode into town,
Into the fastness of Doublethink,
He thought the sheriff quite intelligent.

'Reifying' is a Marxist word for the process
Whereby insiders reassure themselves
That an untenable state of affairs
Is a God-given law of Nature,
Part of the cosmic or genetic fabric,
When a little iconoclasm could change things

For the better, smashing the brick-moulds
Of conventional constraints on thinking.
'What conflict between Custom and Nature?'
Asks the relativist, hardly a Utopian,
'Each to his own conformity; judge not.'
But the sheriff is no ultra-conservative.
He knows that Custom is walled against Nature
And is always aware of the polarity.
He knows the spirit's malady,
He is squint-eyed with the 'maladjustment'
Of individuals to society, but is secretly
Relieved that people are unwilling
To accept society as they find it.

The outlaw brings the individual's maladjustment
Before his sole tribunal of society's maladjustment –
As if Custom could not foster Nature
(As if parents, teachers, vocational guidance counsellors,
Job workshops, friends, preachers and psychoanalysts
Were all in collusion against his nature);
As if every practice were too rigid to be changed by Nature
(Too remote from ballot-, soap- or suggestion-boxes);
As if rules could not be bent and collective action
To change laws and institutions were unthinkable.
'All they who are not integrated
Shall henceforth be called 'alienated'
By law!' – not the law of Nature
But the law of historical inevitability,
Entailing its own precepts, which –
Though less universally understood
And more rigidly and selectively applied
To discourage second thoughts –
Once accepted, after a while,
Will become second nature
And neat as bricks and mortar.

Surely it is natural –
Not in the reifying sense –
For the young, thin-skinned intellectual

To befriend a society with leprosy,
Even as the unnatural opportunist
Picks over its uncleanness for profit,
Reducing the body to numbness and deformity?
The outlaw would administer holistic medicine
For society's atomization
And thickening of nerve,
The capitalist atomizing medicine
For society's collectivity
And finer feelings.
When the Wall came down,
Millions more went up:
Thickening walls of skin around egos,
Corrupting all who touched them,
Augmented by walls of stereo sound,
Comfort blankets of Virtual Reality
And the latest product of interactive technology
(Warning of tomorrow's decentralized,
Entrepreneurial, electronic market-place)
The Being There Wall.
The Athenians had a word
For a person with nothing much upstairs
Pertaining to a social conscience,
'Idiot',
Because it's by interacting with others –
In person, not in cyberspace –
That we construct our reality
And know what reasons to use
To justify our actions.
As our civic ideals are,
So shall we shine
And not be disfigured.

Children can be models of humanitarian concern.
But as society breaks down case by case,
By gender and marital status into voting patterns,
Into the drugged units of profit and consumption,
Its fragmentation sucks their childish crania,

While the white flocks, colonizing the cliffs
Of public discourse, reducing it to beastly syllables,
Peck their green and moral bones clean.
Maddened by society's consumptive disease –
And now with that instilled, modern rationality
That declaims loudly against myth and superstition
And any hint of 'indoctrination' in education –
Man-as-beast will become perfectly enlightened,
The enlightened perfectly beastly –
As it begins to eat its own children.

The death-dealing atomizer was Fat Man.
He supports the fissionable, nuclear family.

Harmless, by contrast, was the Romantic individualist
Who admired Nature in her purity
Or red in tooth and claw
And scorned the strait-jacket, Convention.
Nietzsche, sadly, too unconventional,
Ended his days in one,
While Goethe, the idealist, had integrated himself
As a privy councillor, anatomist and botanist
And kept his head, as did Wordsworth,
A sub-postmaster. Neither would have said
That the society of their day was natural;
Both would have disagreed with some of its customs
And might have tried to change them from within;
But they felt it important, as individuals,
To be comfortable in society,
To feel at one with other people,
Because social interaction
Gave them their individual identities
And preserved their sense of reality and proportion.

But modern man sees himself as *Homo economicus*,
As a sovereign, but faceless consumer or producer,
Or as *Homo scientificus/religiosus*
To be valued only as a confirmer of his theories/dogmas
Or as a bearer of genetic/sinful inheritance,

Rather than as Social Man, the prey of fear and self-doubt.
Yet from where else but our social experience
Do we learn who we are, and how to grow,
How to hope, rather than despair,
How to give, rather than take,
How to find acceptance, rather than feel guilt,
How to be confident, rather than shy,
How to appreciate, rather than condemn,
How to like ourselves, instead of disapproving of others,
How to be patient, instead of ranting,
How to prize justice, instead of being unfair,
How to be creative, instead of destructive,
How to make peace, instead of war,
How to have faith, despite feeling insecure,
How to find friendship and love
Despite the coldness and ingratitude we meet with?

But if society is the agriculture of the mind,
Her guardians have a special duty to respect the seed corn –
Our life, dignity, individuality, autonomy, gregariousness,
Territoriality, curiosity, rationality, creativity and spirituality –
The ten-fold root of our essential humanity.
To be social is our nature and our destiny:
Our minds can accept the need for convention,
For status and roles, institutions and rules.
But we are not clay to be moulded by society,
Or extruded by biological imperatives or psychological drives,
Like characters in a play we did not write.
Ideally, we decide how we shall act out our desires
Within the rules, how we shall form our characters,
What roles in life to adopt to express
The character which is our holy uniqueness
And the humanity that is our redeeming essence,
Decide the degree of our assent to market forces,
Our closeness of fit to the experts' characterization –
Decide how we shall influence the relations between people,
Whether to enrich or impoverish,
Whether to turn the other cheek or turn a blind eye,
How we shall leave our mark on the world.

But the rational disposition of the freeman will be limited
By ignorance or powerlessness, prejudice or poverty,
By bribes, manipulative parents or politicians,
By malice or greed, unskilfulness or bugs in the software,
By suggestive ads, by propaganda or blind religious affiliation,
By drugs or technologically-induced artificial experiences,
By faith in scientific progress, outstripping faith in social.
These limitations will make us ham actors,
Forcing the few lines that life has given us
Into a thousand, untimely, stilted utterances.
The utterances are ours still,
The characters ours,
The mannerisms ours –
But only if they complement limitation with affirmation
In deference to reason
And in deference to our own true motives –
To the real people we are
And to the real needs of people.
Then, and only then, shall we be
Actors in the drama of our own lives,
Speaking our own lines and adopting characters,
Instead of having them thrust upon us.

Just as a plant cannot develop its nature
Fully until it finds its own place
In deep, fertile, weed-free soil
And it has watered and sunned itself,
So if our institutions, laws and customs
Are inhumane, if our culture and *mores*,
If the play is idiotic – nurture will restrict
Ordinary nature to a spindly growth.
Superficially, our moral maturity will wither,
So we justify less our social behaviour –
Which can happen to the best of us,
For example, the breed of managers and consumers.
In dusty soil essentially human roots shrivel.
This corrodes the brain, so that we cannot
Confidently give genuine reasons for our actions
In terms of the roles that we choose to play,

The norms that we value, the shared script,
The views, feelings and aspirations of others,
While having regard to the immorality of greed,
The need to protect the vulnerable from harm
As well as the limitations of human knowledge.
Such reasonable speech grows less evident
And, in so far as the rational is real,
The characters we play become less real.
Then those nice men in white coats will come
To make us act in the ways prescribed,
To drug us with statistics, economics, cybernetics
So that, bleary-eyed, we ask
'Will it make money?'
'Will it work?'
'Will it receive corporate backing?',
Not 'Is this what society needs?',
'Will I be happy doing this?',
'Is this really me?' –
Questions no self-respecting atom would ask.

VIII

I, Citizen

(In memoriam Kevin Jones, who taught me the Elemental Analyzer.)

'Society in its full sense . . . is never an entity separable from the individuals who compose it. No individual can arrive even at the threshold of his potentialities without a culture in which he participates. Conversely, no civilisation has in it any element which in the last analysis is not the contribution of an individual.'
– Ruth Benedict

1.

'No individual can arrive even at the threshold of his potentialities without a culture.'

'No individual can arrive even at the threshold of her potentialities . . .'
If eugenics and gene research do not integrate,
If drugs and chip implants are not used to stimulate
The brain-waves we scan for their potential, so great
To serve the State.

'No individual can arrive even at the threshold of his potentialities . . .'
But for the commerce in which we participate,
Expanding the markets to cover his plate,
Developing the technologies to fill his pate
Outside our gate.

'No individual can arrive even at the threshold of her potentialities . . .'
Till she comes to the enclaves of peace and prosperity,
Those havens of culture and the cutting-edge company,
Walled against that visitor from the dens of austerity,
The unsuccessful interviewee.

'No individual can arrive even at the threshold of his potentialities . . .'
Until he has learned the true value of pelf,
Retrained to avoid being put on the shelf,
Jockeyed for promotion – so vulnerable that elf
When left to himself.

'No individual can arrive even at the threshold of her potentialities . . .'
Till cultural tradition has defined what is normal,
Made the intelligible boundaries of experience less formal
With a diet of drugs, TV and the paranormal
To sustain her morale.

'No individual can arrive even at the threshold of his potentialities . . .'
If he is exploited, diseased or depressed;
But he regains the capacity, when he's convalesced,
To justify his pay, knowing it is the best
Endurance test.

'No individual can arrive even at the threshold of her potentialities . . .'
Till she understands what her culture expects of her
And tempers her curiosity to her role as babysitter,
Her desire to exert influence to demands on the ironer,
Unlike her teacher.

'No individual can arrive even at the threshold of his potentialities . . .'
Till he makes informed choices about how to grow,
Has confidence to act so that others will know,
Answers for his failures and successes, although
Not the *status quo*.

'No individual can arrive even at the threshold of her potentialities . . .'
Till she can defend not only what she does
But the person she is by the dignifying buzz
Of her commitment to others, or to a cause that was
Opposed to us.

'No individual can arrive even at the threshold of his potentialities . . .'
If the needs are unsatisfied, that make all flesh kin,
For health and autonomy, to be centre not margin;
But to rally the oppressed we shouldn't begin –
Sleeping dogs within!

2.

'Every element in a civilization is, in the last analysis, the contribution of an individual.'

'Every element in a civilization is, in the last analysis, the contribution of . . .'
Ideas ricocheting around pinball brains,
Electrochemically flipped, as ions through membranes,
Till nought of ethical decision-making remains
In our cerebral chicanes.

'Every element in a civilization is, in the last analysis, the contribution of . . .'
Defenders of soul and stalwart individualism,
Who regard the personal as incompatible with mechanism,
Who put commerce off limits to scientific reductionism –
To fulfil its own cynicism.

'Every element in a civilization is, in the last analysis, the contribution of . . .'
Technological breakthroughs, transforming life and economy,
First the Industrial Revolution, then the microchip hegemony:
Negotiable in these markets any coin with the imagery
Defaced, of humanity.

'Every element in a civilization is, in the last analysis, the contribution of . . .'
Individuals prophesying changes in theology,
In style, demand, competition or technology,
Or catalysing those changes, innovating a psychology
With a restless pathology.

'Every element in a civilization is, in the last analysis, the contribution of . . .'
The rare who cash in on their individuality,
The larger-than-life star on field and TV;
More coercive is the despot, more liberating the luminary
As objects of idolatry.

'Every element in a civilization is, in the last analysis, the contribution of . . .'
The individual personality, moulded by convention,
By the religious, political and occupational dimension –
The 'framework' of suicides that to Durkheim meant suspension
Of empathic comprehension.

'Every element in a civilization is, in the last analysis, the contribution of . . .'
Individuals one so treats by understanding their needs
And personality, so one speaks as tactfully as heeds:
The diplomat in family or society reads
Individuals, not breeds.

'Every element in a civilization is, in the last analysis, the contribution of . . .'
Individuals who still need recognition by another,
Whether captains of their destiny, or toadies to Big Brother:
Even the captain, self-presenting, his private thoughts would smother
In their beds – like Big Brother.

'Every element in a civilization is, in the last analysis, the contribution of . . .'
Machiavellian manipulators, not empathic helpers:
Without ideological burdens, they're utilitarian achievers –
How they evaluate the needs and potentialities of others,
These empathic self-helpers!

'Every element in a civilization is, in the last analysis, the contribution of . . .'
Us with the power to get things done,
Mac stuffing the faces, every one,
Lest they should grow to Twenty-one, –
While millions have none!

IX

Operation Resocialization

'Children mothered by the street . . .
Blossoms of humanity!
Poor soiled blossoms in the dust!'
(from The Street-Children's Dance (1881) by Mathilde Blind)

(This poem is dedicated to the memory of the eight who died in the Almaty camp, within the walls of rejection and helplessness thrown up by adult indifference, and to those young people, numerous as the stars, who have known only those walls – or no walls at all.)

We, the peoples of the United Nations determined[1] . . .
'Nobody invited you here, boys,[2]
You came by yourselves . . .'
To affirm faith in fundamental human rights,
'Here you are in a prison camp,
Some may call it a pioneer camp . . .'
In the dignity and worth of the human person . . .
'It's well protected . . . there are guards
At every watchtower, armed with live ammunition . . .'
The Universal Declaration . . .
'We gave back everything that we'd stolen,
Except the ice-cream, which we ate in one go . . .'
Of Human Rights, Article . . .
'The sentence was four years . . .
One, two, three, four . . . One, two, three, four . . .'
Five. No one shall be subjected to . . .
'You take a needle and dip it in petrol,
You stick the point in your leg. The next day . . .'
Torture . . .
'Your leg swells up and it hurts so much you can't walk –
The good thing is that nobody knows you did it yourself . . .'
Or to cruel, inhuman . . .
'Anyone who makes a noise

Gets his head kicked in! . . .'
Or degrading treatment or punishment.
'I want the whole floor
Licked clean till it shines! . . .'
Article 25. (1) Everyone has the right . . .
'A 14-year-old boy was sent to prison, he spent
Two years there, then they told him he was free . . .'
To a standard of living adequate for the health . . .
'He said, "There's nothing for me out there,
Things weren't so bad here" . . .'
And well-being of himself and of his family, including food . . .
'I get fed three times a day,
I have a bed, what more do I need? . . .'
Clothing, housing and medical care . . .
'I won't freeze to death somewhere because I'm drunk,
Because if I do get drunk here . . .'
And necessary social services, and the right to security . . .
'They just put me in the cooler
And there are bunks there. So everything's fine . . .'
In the event of unemployment, sickness, disability, widowhood, old age . . .
'Some old grandma fell
And people just trampled over her . . .'
Or other lack of livelihood in circumstances beyond his control.
'People get used to anything,
Humans are like that . . .'
(2) Motherhood and childhood are entitled to special care . . .
'What is a leader?
He looks after the boys in his gang . . .'
And assistance . . .
'But to take care of his gang,
He has to put down others . . .'
The child in care has a right to individual respect and consideration . . .
'For example, he might help me get more clothes
Which means taking them off somebody else . . .'
A right to be looked after by skilled adults who have a commitment . . .
'So someone ends up
Without their belongings . . .'
To the understanding and meeting of his individual needs . . .
'He takes clothes off the weakest –

It goes on down through the ranks . . .'
An assessment of his family and home environment should be undertaken . . .
'Until you get to the very bottom,
The person at the very bottom goes about . . .'
The child in care has a right to live in an environment . . .
'In rags,
Have you seen them? . . .'
Conducive to his emotional, physical, social and intellectual development . . .
'They're the ones at the very end of the line,
With nowhere to go . . .'
The child in care has a right to individual attention . . .
'Give him a rope and he'll use it to top himself –
That's how these boys, called crapbags, are born . . .'
Which shows recognition of and respect for his unique identity . . .
'Name of patient: Plotnikov, Zhenya, 16-years-old . . .
"I swallowed a cross of wire (trussed, worming itself into bread)" . . .'
A right to information concerning his circumstances . . .
'Diagnosis: foreign body in oesophagus . . .
People outside have stopped being people . . .'
And to participate in the planning of his future . . .
'O my Lord, save my sinful soul
From local punishments, from the far-away zone . . .'
The child in care has a right to administrative standards . . .
'Being frisked, the tall fence,
The severe public prosecutor, . . .'
And procedures within his caring agency ...
'The devil owner, small rations,
Prison housekeepers, steel handcuffs ...'
Which will protect him . . .
'From hard
Labour . . .'
And promote his interests . . .
'From a cold cell
And short haircuts . . .'
The child in care has a right to the protection of the law.
'Save us from the death penalty.
Amen.'

[1] *From the UN Declaration of Human Rights and 'Children in Care: A BASW Charter of Rights'*

[2] *Excerpts are from Experiment of the Cross, a TV report (C4; 30/7/96) about corrective youth camps in Kazakhstan, based on clandestine footage of one camp at Almaty obtained by a shocked psychiatric doctor, Dr Taras Popov. Internees, many of whom had committed only petty crimes, were subject to a brutal punishment regime. Disease and sexual abuse were rife, and suicide common. The inmates' self-inflicted injuries were attempts to escape the gulag for the relative comfort of hospital; it backfired in Zhenya's case (he hoped that the cross would open up in his stomach, causing limited damage). Since the first screening in Russia in 1995 (which cost Dr Popov his job), 8 of the inmates have died: 5 died from TB, hepatitis and malnutrition; 3 were murdered. Amnesty International is now campaigning for an investigation into these deaths and for the reform of such prisons in the former Soviet Union.*

X

In the Light of Almaty

Hard-liner: In the light of Almaty
Now says the smarty
That penalties should be humaner;
But then the hard core
Will steal all the more,
So how is your soft-line saner?

Soft-liner: Grows our morality
From first fear of penalty
To the maturity of ethical reasoning;
From the 'No!' of parents
To our fellows' forbearance,
It adapts us to communal living.

Hard-liner: What stabilizes a family,
A company or society,
Is compatibility of members' aims:
A guarantor of co-operation
Is enforcement, not advocation,
A tight ship, a morality of shames.

Soft-liner: A galley of slaves
Could bring you your raves –
Will their duty not try your conscience?
Or will respect for convention
Blind you to the distension
Of avoidable suffering with affluence?

Hard-liner: No duty is owed
To a higher code
To question poverty's persistence;
Hence crime is defined
In the public mind
By sanctions, not social systems.

Soft-liner: Food is destroyed
So that prices are buoyed
While millions are subjected to hunger;
Is the public wrong grosser
By the pusher or doser,
By the cardsharp or insider dealer?

Hard-liner: The punishment of business
Or political remissness
Is an alternative more rhetoric than practical;
Its consequences ramify
Too unforeseeably to justify
Your reasoning that capitalism's immoral.

Soft-liner: You look for consequence
To argue the expedience
Of doing right in pounds of self-seeking;
You then call 'unreasonable'
That right, termed 'desirable',
To universal economic well-being!

Hard-liner: If the principle of equality
Had greater fecundity,
I'd consider my position dogmatic;
But it's taken for granted
That Marxists recanted
And our society's unproblematic.

Soft-liner: Is the system reasonable
If we choose untenable
Solutions that damage our interests?
For ghetto immiseration
Fills the police station
And Third World pollution's unaddressed.

Hard-liner: If people were rational
They'd be socialist or criminal
In a grossly unequal society;
But they prefer normality,
Not guided by rationality
But a stronger leaning to *conformity*.

Soft-liner: Habit and consensus
Won't lower the fences,
And *justice* is a reasonable standard
For shaping institutions
And reaching decisions,
Since far from equity we meandered.

Hard-liner: Why talk of justice?
Having stations and duties
Brings stability, order, tranquillity,
Three rights that are innate
To the human state
And basic to the sentiment for property.

Soft-liner: Duties and stations
Won't heal the nations
Or realize human potential,
Till citizens' aims,
Grievances and shames
Are to the morality of all precedential.

Hard-liner: That is a recipe
For encouraging every hippy
To criticize and modify institutions,
Instead of adapting
To the principles they're wrapped in
And to the aims of corporate organizations.

Soft-liner: Is humanity an end
Or the means to an end?
Human rights create orderly conditions,
Where greater equality
Brings mankind more dignity
And enthusiasm for collective decisions.

XI

Radical

'The sensitive plant has no bright flower;
Radiance and odour are not its dower;
It loves, even like Love, its deep heart is full,
It desires what it has not, the beautiful!'
(from The Sensitive Plant, by Percy Bysshe Shelley)

INTRODUCTION

Our nature is a portrait of balance and proportion:
A head not too big for the shoulders, or shoulders
Too big for the head. Ugly is our portion
If gems in our nature become little boulders
Or the society gardener stints the powers
Of our roots to sustain the early rose that flowers,
Blighting our idealism, till the whole bush moulders.

The core of our being is no secret spring,
To be sought by scientists, though deep in the ground:
There's a taproot, life-guzzling, with nine side-roots issuing
That feed that thirst and anchor it sound
In social reality. How roughly the boot
Must heel in the bush, to disintegrate this root!
How laterals dilate in waterlogged ground!

1.

LIFE

(the zest for living develops into the primary root)

'Is it true that grown-ups have a more difficult time here than we do? No. I know it isn't. Older people have formed their opinions about everything, and don't waver before they act. It's twice as hard for us young ones to hold our ground and maintain our opinions, in a time when all ideals are being shattered and destroyed, when people are showing their worst side, and do not know whether to believe in truth, right and God ... It's really a wonder that I haven't dropped all my ideals because they seem so absurd and impossible to carry out. Yet, I keep them, because in spite of everything I still believe that people are really good at heart.'

(from The Diary of Anne Frank, 15 July 1944)

'Suicide Mother Kills Sons in Frenzied Attack.'
'Uniform Approach to New Ulster Boycott
- David Sharrock on the Catholic parents, shunning a Protestant shop,
possibly leaving pupils without clothes for school.'

(from headlines in The Guardian, 29 August 1996)

The sap that rises in young hearts of oak
Oft smells of roses, so instinct with life
That it buds in the Fall. Human spirit never spoke
From its centre more movingly than Anne Frank from the strife
Of uprootings, crude enough for the metaphor. Fully found
Was the root of her humanity, integrated albeit pot-bound,
Till transplanted and heeled in, and out, of this life.

She imbues me with a sense of life so precious
That she lives, while the living are the walking dead –
Like the religious obsessive, beset by her incubus,
Or parents who use children as vehicles for hatred.
The virtue of the taproot at the core of personalities
Is its wholesome control of traffic from the extremities
With its priority command to *Give Way Ahead.*

As laterals probe the earth to extend the psyche,
The taproot holds the balance between growth and standstill:
A policeman on point-duty isn't perfect, but by crikey

Without him to prioritize, traffic jams beyond retrieval,
We're overrun by heavy goods, bandwagons, sleepwalkers,
Our homes and streets rumble, we fear the night stalkers,
The goblins of primal sin that caused Mother's upheaval.

The fanatical fuse blows sky-high the head;
But where the boot goes in or the land is ungrateful,
Will pride-leaching poverty keep the proportions fed,
Or will daughters be sold to a pimp for a plateful?
More blameworthy are the well-watered with lives out of proportion
To the law of human nature, who connive at abortion,
The murder of innocents by Hitler's new faithful.

The human taproot should sensitize the rest
To the value of life, being stronger than side-roots
That know only consistency. Do we act for the best,
Harmonizing root needs, resolving their disputes,
Trying to impose one structure of feeling
That life is a show no lateral should be stealing,
But all play their parts in feeding the shoots?

Though a bad piece of planting won't leave the root feeling
That existence is good, and the more of it the better,
If we cherished each life for its uniqueness, repealing
The self-justifying price-tag and standard criteria,
We'd replace the nurseryman with a nature lover
Who limes the holes, since we'd all be poorer
Minus the rose-bush whose potential was fettered.

If we care for the bush, we'll care for the roots.
Each lateral fits somewhere into the background of a life,
Each strand of the life force. Moderation, that suits
Survival, wisdom and keeping a wife,
Is the taproot's function: it flows with goodwill
And reverence for life – if the soil is not ill
With the warlike canker in taproots so rife.

2.

DIGNITY

(the latericumbent)

'In his private heart no man much respects himself.' – Mark Twain

'Gloire de L'Homme' has a well-proportioned root,
Each lateral like a mainspring, an atavistic impetus
That makes us tick. All converge at the taproot
In a balance of power, in a poise for wholeness
Between conflicting drives. If we lack equanimity,
We might have suffered damage to our personal dignity –
To which 'Gloire' is vulnerable when the soil is pressed.

Poor trampled dignity! A strand convulses
In the smeared medium, soured by the press
And advertisers' commandeering a rout of impulses.
Now over-stimulation foments unrest,
As rationality, subtlety and loftiness of motivation
Are disavowed by the worker, the shopper and nation.
Now high-principled Job gets his knickers in a twist!

The child may hide talents in a head not bitten off,
The adult may be redeemed by qualities unrecognized;
Others keep their dignity if they are not written off
As unmarketable goods, superficially apprized.
The justice that is done to our full capability
Brings the acquittal of ourselves with commensurate nobility,
Unlocked from the cell of ourselves, the baptized.

If people were deemed as inscrutable as they are,
Not pigeon-holed, homogenized or reduced to a gene;
If the handicapped weren't shunned, kids belittled; if the bourgeois
Weren't exploiters and workers commodities; if a machine
Could give freedom; if strangers were given welcome – humanity
Would reclaim what egoism begrudges, a dignity
As characteristic as affection is the recognizer's mien.

3.
INDIVIDUALITY

(the unilateral)

'Whatever crushes individuality is despotism, by whatever name it may be called.' – J. S. Mill

The cultured loam, the breath of roses,
Root needs poised on the edge of a knife
Is the metaphor I choose for the idea it proposes,
That I only find myself by losing my life
In what I share with others. By devotion to a group,
Able to call forth an aspirational whoop,
The unilateral thrives in the soil of life.

But the soil that grows me may also suppress me.
When acidity is critical, some cultivars are susceptible
To isolates of the unholy *Zeitgeist nietzschi*:
With their strong individuality they'll make me suggestible,
An imitative one among a faceless many.
As the power of demagogues may lose me my dignity,
So mass consumerism will regiment the gullible

With its carrot-dangling style and pandemic persuasion
Drowning discretion in soma-pop and cola.
Another form of standardized, cultural communication
(The limits for taste being set by the dollar)
Sows expectations of gangs, bangs and mayhem
In the viewer, who finds in the funny farm asylum
From his own true need to be programme controller.

Individuals are persons who know their own minds
And act in that consciousness, the natural estate.
Could we seek again the dark thickets of our minds,
Send the apes packing, what could we not donate
To the life of society? But farewell individuality,
Farewell the hope of good things from originality,
If the oil for the cogs is the standardized lightweight!

4.
AUTONOMY

(the assumed equilateral)

'By a careful cultural design, we control not the final behaviour, but the inclination to behave – the motives, the desires, the wishes . . . we increase the feeling of freedom.'

– B. F. Skinner

A rose expresses her virtue, and sighs,
As the breeze stirs a pang of fertile earth's essence,
So cultured a pregnancy that the soul must rise
From its social roots with the sap's conductance
Of fine reasons and choices; those she missed
Drain back to the soil – her fertility unexpressed –
Where the water, now pooling, reflects blue expanse.

Society's sickroom is oppressively still.
With no interest in mending, we shout from the window
'It's our nature to be selfish, vengeful – ill!'
What glibness denies a free agent may show
And we slaves, socialized with the semblance of freedom
To avow the all-too-human, in sickness become
Neglectful of Number One – whom so few of us know.

For no less an individuality than the one we aspire to
Can be ours! This bourne is a land of orchards,
Of juicy plums, and free spirits who desire to
Trample the husk that no kernel guards!
Now let's pick values! Let these fill a basket:
Self-interest and altruism, kindness and grit,
Profit and conscience, science and bards,

Freedom and order, honour and love!
Try to prevent them from turning to jam!
Is there any contender that should be written off?
Do we balance the claims, or not give a damn?
Are we radical amputees, or centred in plurality?
Can we say from the heart of our individuality
'Our nature is to weigh, not excuse *ad nauseam*'?

5.

GREGARIOUSNESS

(the collateral)

'What the crowd requires is mediocrity of the highest order.' – Preault

The rose does not manifest the clod, or dedicate
Her life to the soil. She enjoys being a rose!
From humanity's cheek flew the roses of late,
When the collective conscience the private froze,
When the many could bloom the one out of countenance
And drain individuality, till all the significance,
Life and breath of society were shows.

Society lives only in the cement of my person,
In the clay and lime it must grind and roast
For mixing with the sand and sea of my halcyon,
Expansive days. And the more I grossed,
The more I owe, the more life I can give
To society. Then what gives the docile and imitative
Swarm? Continuity and belonging to the rose!

The rose courts no roseate but a stripe-suited air marshal
To shake her gold hands: so the hive needs true
Outgoing sympathy, proboscis-impartial
Towards blooms and buds of all hues. Virtue
Oft on the wings of pollinators has flown
To blooms not aggressing or ingratiating to be known,
Whose nectar is not sickly but a bee-welcome brew.

But greed is gilding a mantrap, to estrange them
From the buzz of humanizing encounters, their womb.
Like mooning wolf-children, the feral will derange them,
Having torn their humanity from the threaded loom.
That tragic nonperson, Hauser, did not thrive:
He cherished inanimate objects as if live,
Thought like the babes that Avarice will groom.

6.

TERRITORIALITY

(the laterinerved)

'The moment we care for anything deeply, the world – that is, all the other miscellaneous interests – becomes our enemy.' – G. K. Chesterton

Llandudno beach is soaked with crowds,
Pock-marked with huddles, peppered with somebodies:
Dominating space, groups bunch like clouds,
Units protective of the ties each embodies.
So family and true friends beautify the possessive
Reflex, and even, fearing strangers, the aggressive –
While the girl not defending her space is anybody's.

Who gains my trust I take in to my redoubt,
But I disown the abuser – like the defaulting borrower
And her sister, mocking me who bailed her out.
The bane of good neighbourliness is manifold: the drug-pusher,
The corrupt, the unruly, the thief, the parasitical,
The nit-picker, the gossip, the wife-stealer, the vandal
Of real roots and sympathies are the overthrower.

Trust and property are the markers of territory,
Making kin our kingdom, complete with a castle.
But trust must be honoured or property miscarry,
And the territorial imperative is no Band-Aid facile
For a stultifying home-life, workplace or nation,
Not marching towards human fulfilment and toleration,
Not fostering mutuality instead of the vassal.

The protective and selfish both defend their space:
One fights to win affection, the other to gain respect.
But protection has its limits and selfishness its place,
If the way of self-fulfilment we are not to neglect.
We look after our own, in the good sense and the derogative,
Hating change, intrusion and threats to prerogative,
Fighting now for love, now for respect.

7.
CURIOSITY

(the plurilateral)

'A fool has no dialogue with himself, the first thought carries him, without the reply of a second.' – Marquis of Halifax

Natural is the marking and defence of territory
With stigmata, strata and specialist data.
But equally natural is the intellectual artillery
That mows down divisions and disarms the Tartar
(Defensively sniping from delusion's thicket
Of religious, social and scientific etiquette)
By showing him ways that he could be smarter.

Inquisitiveness is the urge very few enlist
To guide and control the others. Too subversive,
It confounds alike the sceptic and reductionist,
The clone, the little islander, the dogmatist and coercive.
Sycophant or questioner, which is the feebler?
'My brain for outspokenness,' crowed the *enfant terrible*
'Is better suited than yon Emperor in his regalia putative!'

Old heads, forgetful of science so captivating
As wonder's fruitage beneath juvenile skies,
Cannot say how or why joy was waiting
Or what surprise gives or analysis denies
Of the felicity they sought then in nature and lives –
But somehow in the field a deep receptiveness revives
To what made their home and bright their eyes.

From the mammae of the future the quick young survivors
Milk knowledge no profiteers can gamble into the song
Of instant gratification they tidy into ledgers,
Hugging every frailty to breaking erelong.
Speculators, who really give profit its place,
Insist on risk-taking for the benefit of the race,
So uniquely curious, so uniquely wrong.

8.

RATIONALITY

(the thinking lateral)

'You can only find truth with logic if you have already found truth without it.'

– G. K. Chesterton

The crown catches fire from the heart's sap – the individual
Fervid thinker crushes the ferret –
The snow-covered mountain is majestic but cruel:
The rationalist must decide if lives matter and have merit,
Being more than statistics. Unless he sends
To know for whom the bell tolls, that portends
His mortality, its clang will din into him his demerit.

Practical self-interest is rubbing its cold hands
At its rational inheritance, after its prioritizing heart
Was buried at Wounded Knee and in the airy-fairy lands.
Now Reason, disembodied from its throbbing counterpart,
Computes the means with the cost and the benefit,
Dropping the still, small voice as a prerequisite
For making moral choices. It thinks itself smart

With its one cold eye on profit! The pupils
Of the Cyclops, with their feet on the ground, are resplendent
With the utility gain from reducing their scruples.
They sum their endowment of Reason to be independent
Of the way they live, ever stamping their feet,
Uprooted from the soil of otherness, the seat
Of thoughtful responsiveness. Surely pendent

In air is the taproot, if shrivelled are the priorities
At the centre of life, if unbalanced are desires
By the mind? Is the plant stressed? It is!
Sans heart, *sans* values, mind quenches its fires
Till it knows what's what, nor cares what's not,
A student of means, with the ends forgot –
To be whole, to have character, to know what inspires.

9. CREATIVITY

(the lateral thinking)

'No matter how old you get, if you can keep the desire to be creative, you're keeping the man-child alive.' – John Cassavetes

This is self-knowledge, to shut out the 'creativity'
That screams on all sides to sunder us, invitations
To consume – to eat up our self-sufficiency,
Feeding a false sense of our own privations.
Self-knowledge is rebuking Reason's illusions
That growth and technology are creative solutions
To discontent. Self-knowledge is the mother of creations.

Cold Reason makes capital: Creativity makes friends,
Makes music and dance, makes up, makes love,
Makes laughter, makes believe and talk that never ends,
Makes artful religion from the ceremonials thereof,
Ready for painting with brush or pen,
With the strokes of desire, that animating oxygen
That fills out our canvas and the sheeted ladylove.

Not limping now is Long John Silver,
As when insight comes one hop at a time
To the bureaucratic and scientific problem-solver,
Hobbled by system and hypnotic paradigm
Into plodding re-analysis of what he already knows.
Breaking in my head are the shore's counterblows
To the inhuman pattern, exposing the pantomime.

The jess that is tied to the untameable falcon
Becomes jest after a while. A dove undoctrinaire
Alights on the landlubber's shoulder from a galleon
On the immense illogic, and cries, 'Ahoy there,
Me arties! Mind, take off your clothes!
Eyes, be refreshed to juxtapose
Incongruities in the look and manner of the square!'

10.
SPIRITUALITY

(the idolateral)

'The facts of life do not penetrate to the sphere in which our beliefs are cherished; they did not engender those beliefs, and they are powerless to destroy them.

– Proust

From a symphony of motives at the heart of thinking
A beauty resonates which is a confirmation
That breath is sweet and the root is drinking
Deep from the mutual and enriching relations
In which it is grounded, and the degree of its permeance,
As it gently takes and gives of its substance,
Informs every one of its lateral affirmations. –

It's not the reality, but the ideal I'm proposing!
I'm still pressing on in the direction of wholeness:
If only the barrier would crumble before the nosing
Root-tip; if only the dull mould of soleness
Would simplify and yield to more openness to others –
My radicalism would go more with the grain of the universe,
Loving and marvelling, well-rounded in boldness.

For something there is in the experience of becoming
More human that enhances the mystery of man
And human experience, and makes it something
To hold onto, to believe in, to build on, if we can;
And if, by being invoked, it inspires us to grow,
It provides food for our hope no science can overthrow
And the mystic will finish what the reductionist began.

The visionary eye is most akin to the amorous,
Though cosmic in its range. Now it prefers Nature,
Excellent and true, to the man-made and spurious,
To bear sons of wonder, in her arms secure.
The visionary eye is not narrow and supercilious,
Disdaining surprise; nor vacant and credulous,
Affecting escape: it is wide with adventure!

CONCLUSION

'Know then thyself, presume not God to scan;
The proper study of mankind is man.'
– Alexander Pope

The road to ruin is the defeat of aspiration,
The curtailing of humanity by unemployment. But few men
Think work or prosperity could cripple a nation
Too skilled to remember what it means to be human,
Too busy to be interested in becoming more so.
The validation of our humanity is not a *quid pro quo*
But for hearts of gold, by the question driven

'Who bade the rose-bush to sculpt her bed?'
If you trust not the poetry, consider your sanity
Which bids you to anticipate and follow the thread
Of your actions on others. All else is vanity,
But for the intuition that love is cosmic
And that 'Gloire de L'Homme' has the one thing basic,
The tenfold root of a common humanity.

BOOK 3

Attempts to answer the question:

'Who Am I?'

'Be a friend to thyself
and others will be so too.'
– Proverb

October – November 1996

I

Mr. Mee

Thrill-seeking child, your carriage awaits your life!
Take the built-in controls
And drive towards skill and self-determination in the strife
'Twixt chance and your journey's goals!
Be agent and experiencer at the wheel – but take your instructor
To reflect on and evaluate your driving and script your roles!

You should know Mr. Mee. He monitors each thought and feeling;
He's the self-aware part of you.
Let him curb and judge the impulsive Ego, repealing
The peer-group pressures that make you
A passenger in life, lest, poorly directed, he loses
His way or his patience with you, and you grate anew.

He will calibrate your speedo; he will set the goals that you long to
Achieve, O would-be examinee,
Set your standards by the ugly or kindred you belong to
And he'll make what they think of thee,
Your abilities and dreams, raise or lower your sights
Till you feel like cancelling the test or proving your proficiency.

Mr. Mee will sanction your emotional reactions to the highway
As your Ego's intuitive friend.
So his responses are the echoes you'll follow when you drive it 'my way'
According to the course he's penned:
His orders will lead you out of the chaos of your feelings
With markers left by you, or for you by others, that he kenned.

Students of the Self find Mee more amenable than the driver
To the behaviourist schemata of labs
Where, seeing only conditioning, they deny the driver;
Ethologists, who like to keep tabs
On impulses and obsessions, deny Mee. Yet 'I' must drive
And Mee must brief, poised the hand that grabs!

II

Commemoration

Below the slate line where Ogwen Bank's parked,
Where the quarry's funereal lava lies,
Is a tumble of sheets – and one was earmarked:
A young holidaymaker, loath to vandalize
With hackneyed graffiti, left real-life scratchings
Of developmental interest to ward off anonymity
And pinpoint his place in the scheme of things
With that gift children have for the nitty-gritty.
The words of the scribe that measured up to his youth?
'KINGSWINFORD 20/9/96 TONY LOST HIS TOOTH.'

I deduce that it speaks of a wound not dealt out –
Of the stubborn nature of flesh, not man
(Which he'd be pleased if more vaguely spelt out
For creeping disillusion's painful scan); –
That openness to feelings was the family piety
And would scarce release a boy in thrall
To himself with 'self-conscious' codes of propriety.
For, self-wrapt, he assumes as common to all
His natural feeling for corporeal history:
Pegs knocked that are calamity or shed that are victory!

But, theming himself, he'd reached introspection:
With a universal tusk he'd engraved a setting
For the jewel that self is. Faceted to perfection
By carving bounds between ourselves and our setting,
Self wanted for no tooth to test the limits
Of his control over events or environment. The plaque
Would fit any tale of striving that licks
The self into shape. That teensy waymark
On our sediment with memory's glancing shard
Commemorates the birth of our self-regard.

When tasty and dry stalked abroad,
We could plot our lives by our hypnotic cries
And telling smiles, for they pulled a cord
Attached out there. But the idler's prize,
Fond reflex of our need, was a Brobdingnagian train
Of routine ministrations, running to time
And unevent, that can breed milksops and drain
Penpushers. When the thwarted and curious climb
To objects out of reach, they leave habit on the shelf.
Some mothers do not encourage this questing for self.

Our independence flowers in unusual locations
And disturbed soils (with florescence of gums,
Apparently!), but fails to bud in situations
Which stir no thoughts of mastery. Life's plums
Are sweet when we face and overcome an obstacle.
Climbing Mount Growth may punch us in the gob,
But just 'TONY' scribbled has no teeth at all!
Such self-advertising does no morale-boosting job,
While the candour that sets numbered teeth on edge
Will find its food in *real* meat and veg.

None of the subtleties to suggest that hardship
Erects a façade: it must falsify experience
To think, legalistically, of dental ownership
Instead of the time and place of accidents.
He learns his good from knowing a neighbourhood,
Close to Stourbridge, with a school, a park,
Friends and a dentist, and the bonds of blood.
Given a clean slate, not marred for a lark,
Tony packs a punch with his sense of 'I am!'
And 'I've a tale to tell!' and 'I'm from Birmingham!'

What appearances he saves by omitting the surname
Are banal, for he writes like one who knows
Where he belongs. Clueless, we frame
A C. V. identity for ourselves that flows
With the commodity stream, an open-grave brochure –
Unglossed by sentiment, the good and bad times

That shape us – the bare bones of a developing structure
In words just as stiff and fetid! It begrimes
Life histories, gesturing obeisance to others,
To clean up the references that make us brothers.

How much are we what our parents approved,
How much what we feel and dream? So unfulfilled
Was my conventional life by their name it behooved
Me to answer to, that I latinized it, my dreams to build.
Renamed, am I anything by virtue of place?
A little grubby-kneed from my boyhood den.
Am I anything by virtue of running in the rat-race?
Still rooted to the starting line by idealism's ken
From ten seasons old. Am I anything by breeding?
None the worse for wear, or better at heeding.

Am I anything orthodoxly? A wandering planet
Or free-floating craft to a star, not reached.
The hallmarks of identity aren't carved in granite:
Paternalistic pigeonholes can be breached
To reflect the complexity of the people we are,
To develop the inner space from which our life grows!
We become aware of the people we are
In the landscape of our very own shaping, in the blows
Of our hammer, in striving our fate to control,
In daring to suggest, daring to be heart-whole –

And daring to engrave a slate with our hand.
Environmental modification is not wanton destruction,
Any more than art is graffiti. The borderland
Between awareness of our human potential and reduction
(Scientific or violent), between creativity and dissolution,
Is a scene of unprecedented human migration
The wrong way. Tony's innocent solution
Was to go the right way. Towards his vocation?
Who knows? But aspiration is born in the cot –
Where it is strangled more often than not.

III

A Dozen Broken Eggs

(Twelve ways to undermine a child's feelings of self-worth)

1. 'Later dear . . .'

The first endangered self
Comes to us
Wide-eyed in her pyjamas,
Wades through a primeval swamp
Of papers,
To meet the petals of our eyes,
Half-closing,
With a query about a school
Project on extinct animals.
Dodo is her name
(After the unwelcome cry she emits) –
A creature destined for certain extinction,
Because it is flightless
And we are always shooing it away.

2. 'Straighter, dear . . .'

The second gets paralysis of both legs.
He is the 'good' boy,
Afraid to wander from the straight and narrow.
The language of his domineering
Elders and betters he knows best,
While his responses and actions
Forsake his own feelings.
'Lift up your heart,

You look *so* smart!'
Was the simple refrain
Of the National Service outfitter
Who sent him to a school for privates
And ordered his short-back-and-sides:
The boy could see the rhyme
But never the reason why his feelings
Should always be so discrepant with his elders'.
But he could not bring himself to say
'My feelings are lying to me' –
Which was all his parents ever wanted him to say –
And so he conceived the wild notion
Of one day writing a psycho thriller
That would somehow change adults' thinking.
A teacher's exhortation
After he left school
And an experience of the transcendent at 19
Were his first real hope
Of getting an education.
'Good' children,
Who manage to ignore their feelings
Almost as well as their parents do,
Who 'do as they are told' –
When told what to wear,
What to watch and listen to,
What to do and what to say,
Where to go and with whom to go –
Learn fast
How not to stand on their own feet,
And especially how not to trust their feelings –
Which is their most important valuing process.
So values,
Which bring a note never so pure to living,
Make no sound
In the private world of children
Constantly being made to feel inadequate.

3. 'Greater, dear . . .'

The third never arrives at selfhood.
No sooner does she reach the brink
Of her picture of herself as adequate
Than parents revise it upwards:
Better that she should not hear 'well done!'
Than stop learning for a second –
Like the Japanese girl I saw
Whom her perfectionistic parents
Had trained for ambulatory cramming –
But not to look out for traffic.
The futuristic hour is coming
When the breeding and teaching of alphas
Could be ritual in a Brave New World,
Where the parents themselves will be
Programmed to soak up the competitive ethos.
They may feel that this scientific potential
Should not be controlled or guided
But followed wherever it may lead. –
They may be the sort of parents
Who are easily led.

Others are surely led by their exasperation.
Now watch the corrosion of kids in Superstores,
The happy consumers' parade ground,
Where constructive guidance and requests are
Bartered for persistent commands and rebukes,
Where the dumb play of dogs is borne out
By the even dumber play of their trainers.
The repulsion of two negative emotions
Can have no positive resultant in the self.
What meagre shivers of encouragement
And praise have to serve these children
To forge into their adult defensive shield!

4. 'Beta, dear . . .'

The fourth eggshell ego
Grows in the shadow of alpha,
Stands on tiptoe
To catch her mother's eye –
And catches the edge of her tongue instead,
Because she does not measure up
To her sibling.
She would hang on her mother's lips more,
If he lived on them less.
He is potency, she mere existence.
How can she rest, find joy or peace
When all virtue and intelligence abide in him?
But she will grow to be unshadowed at last,
When she learns to be unimportant
Or to live in the one crummy light,
Unborrowed from him,
Reflected from her own pallor and emerald eyes.

5. 'Deflate-a-dear'

The fifth green life to be nipped in the bud
Daily dwindles under the lashing rain
Of answers in mockery like
'Don't be stupid!'
Or the cruel unmasking 'You know nothing'
Or worse, 'He thinks he knows everything!'
Or the last word in surly refutations
'You're just like the others!'
The child,
Finding no adequacy in which to rest,
Is something not quite like a person –
He is defeat.
Rendered powerless to answer,
He has entered into a covenant with silence
Not to know the freedom that self-explanation brings,
Not to grow with those spontaneous tonics,

Smiles, faith in justice and reason,
By which the self learns to stand upright.
Perhaps his insubstantial playmate
Was the self he was never allowed to be;
Perhaps, unable to bear his soul's Stygian discouragement,
He had exchanged inward sight for outward –
As millions of viewers do each day, when they conjure
A visible playmate, a more exciting *alter ego,*
To thrill them with the offer of vicarious living.
Yet grown-ups will never learn to meet the morning
Valiantly, realistically, decently or intelligently,
If as children they were trained
To thrum the void of their insufficiency
With their parents' derogatory noises:
'I'm a bad'un, bad'un, BAD'UN!',
'I'm thick, thick, THICK!'

6. 'Castigate-a-dear'

The sixth, his crumpled face
Held up like an offering,
Is told not to cry, for pain
Is as petty as anger or joy,
And he ought by now to have learned
More silent ways of handling his feelings.
In time his hush will sound above punitive outrage
A familiar note of sufferance.
He will never learn through weeping hours
That there was a cluster of grapes
Ready for picking and sharing, some good
He would instantly recognize that he had made
Precarious by his thoughtless deed – but recoverable;
For the one who treads the wine
Thinks so much of his moral anger
That he forces young shoulders to bear a burden
Of guilt greater than the child can understand
Or dissipate by making reparation.
In vain does he shift his heart to a new hold
On tenderness, conscience and reason

When adults take the very steps likely to
Harden him, morally stunt him
And make him neurotic. If hardness
Is stupidity, it is only the adult's own folly
Staring back at him.

Setting against obduracy the possibility of growth,
The reprover summons to his court the voice of complaint –
Not drowning it with his own disgust
Bred of stark anxiety for his moral beliefs –
But ardent for fair compunction to atone,
To stand stiff-backed against that impotence of spirit
Which blots out longing for the outstretched hand.
For the wrongdoer shrinks from,
Sneers at and feels it disgrace to be
His victim – only because he did not find a way
Promptly through wounding to a just measure
Of his ignominy, to confession, fitting penalty,
To restoration of the nightingale and the faded flowers.

7. 'Property allocator, dear . . .'

Seventh in the nest of sorrows is the brood
Of Tarquins, tangent to the family circle,
Squanderer of his people's careers and relationships,
The surest poison of their tastes and interests –
But a wrap green enough for their social functions.
Tarquin is cocooned from mixing with his peers –
There being no Piers – and is told instead
To 'fashion his gait according to his calling' –
But is just too naff to understand those words.

Samantha successfully quarantined her parents
And put out of reach any larger goals
Than earning good bread and admiring their home.
Not feeling unfulfilled by their one hobby, the TV,
They thought that the pressure to achieve at school
And be more outgoing would come from outside
An unedifying centre, empty of stimulation –

As a termly cheque brought her sense of achievement,
Or proof of failure, and snobbery her sense of belonging.
The air of tension and coldness in the clinic, that stared
Her individuality into silence and seemed to say
'All self-expression and spontaneity forbidden',
Followed her home. In her warm Springtime
It may be that she melted her parents, and even
Came physically close to them – but the pleasure
Of being sure of herself and at ease with others
Was a tale soon told, and soonest forgotten.

8. 'Donator, dear . . .'

The eighth dies a different kind of death
From the self mislaid somewhere in Suburbia:
The slow dismantling of pride by indigence
Or that vague sense of social isolation
That scions of the rootless and travellers must feel –
Even though their souls rise vivid as flowers,
Are very hub of the wheel that revolves –
That will revolve.
The donor, deepening his analytical rut,
Lay on the grass outside the Institution,
When a woman and her two small sons –
Travellers in what sense I did not ask –
Sent a ray of hope across the green
That touched my solitude. It was as if
The whole misbegotten world, the shadowy
Tumult of its striving to be, to find the future
In its heart, had shrunk to a request for shekels
From the gentlest and sanest of human futures.
Whatever claim on my pity she owed to misfortune,
Such gentle loving care did she give her boys,
Reading Tolkien to them, while they grew with nature
(One showed me his dear, pet caterpillar), such faith
In their frugal but sublime future – that I blessed
My serendipity and tipped them for a fair wind
To that happy shore where free spirits won't confound us

With their strangeness, nor the solaces of nature,
Imagination and maternal affection invite icy stares.

9. 'Deus Pater, dear . . .'

The ninth, in a basket too loose-woven for eggs,
Pays for her mistakes with drops of blood,
Is told she is unworthy of that look of Christ –
As if Calvary didn't reflect every human weakness,
But spilt milk crucifies. 'Father, forgive them,
For they know not what they do'; and bairns know?
Surely it's the sheep that leads the lamb astray
To learn the vanities that profane its heart?
An instrument of torture is raised, bearing
A Truth so sanguinary it was mighty to overthrow
All pharisaic, codifying and unforgiving priestcraft.
But Innocence pilloried is not the miscarriage preached:
Love now shines in no dark context,
As we lift off the velvet, to beautify our neck,
The gilded rack on which Christ died
(About as tasteful as a pendant of Belsen!).
By the same travesty of sacrificial religion
A feeling takes root in some families that little ones
Are clay that guilt can model into any feature
In the parents' image. Yet love's to be demonstrated
In the lives they lead, the ends they direct,
The egoism and worldly ambition they forsake
Now, not hereafter – and from hymns of the hereafter
The kiddie-seeds of new life are presumed to sprout!
But what child, when he errs, can become eternity?
He cannot be lived by what he cannot even see!
But whatever he lives with, that he learns:
Living with hypocrisy, he learns to be cynical,
But living with sincerity, he learns to find faith;
Living with the Furies, he learns to feel guilt,
But living with acceptance, he learns to find love.

10. 'Waiter, dear?'

The tenth to be prevented from becoming a whole person
Jostled for the right to fumble-buttoned clothes
And a faultless goddess said, 'Let me do that';
Gathered in his hand the voice of his selfhood
And was told, 'No; you will spill it. I shall pour it';
Went into the garden in search of his personhood,
To smother old fears and learn new courage,
And Dame Watchful semaphored to his groping years;
Opened his mouth for a say in his future
And the postie looked for a cake to shove in it;
Opened his ears to neighbourhood children
And the censor tried to stop them with prejudice;
Opened his eyes to family problems,
But no wider circle gave power of awareness
By reflecting its members as public persons
As distinct from self-basted entities in the stewpot;
Opened his mind to the belief that he mattered,
But was shamed and put down before siblings and visitors,
Nor praised or noticed in his attempts to impress; –
Wondered why he, who could think and feel,
Was ignored as a character, but reared as a consumer.

11. 'Desiderata, dear . . .'

The eleventh is raised in a shaky environment,
Because work and marriage have their ups and downs.
Now the welcome home is not there for any
And family life is a few tense hours
Of unfunny home truths and heart-stopping taunts,
When you feel you miss a beat – only to catch it later
From some vented frustration or scrambled complaint.
A misunderstood prey to guilt, you find
You are always in the dock, testifying against yourself:
Your judge has fallen victim to his heart's insecurity
And he suspects your surliness, knows you for a risk.
Trust would only pander to your outrageous demand

To be respected for yourself and grow to fullness.
Surely the money he earns buys your self-respect:
He brings home the bacon with your name written on it –
As you take a language of occupational hazard.

Divorce may strain consensus about child rearing,
Tearing you between parents, as you save what shreds
Of example Dad set you, before Mum left him
To mould in scorn the mother-son relationship –
A cold-war casting in imperious bronze.
Now you live in this one's house,
But in that one's thoughts, as you grow in fitful visits,
Marooned for weeks by the tossing swell.
You wonder why all that should carry weight in a family
Is obliged by law to carry weight outside it
And support the Jezebel who stole his son
And snuggles into her Benefits with a fly-by-night 'dad'.
Paramours and step-dads, who coldly observe you,
You view as rivals for your mother's affection
Or cuckolders of your father – but you save your anger
For more vulnerable siblings, peers and teachers.
As you drink to the dregs the disorienting wine,
You sense the loss of self-controlled lives
Eating deep into your anxious, unpraised soul.

12. 'Traumata, dear ...'

The egg, who's the yucky or fractious attention-seeker
Clinging to teachers because her parents are Teflon
(Either sorry for no wrong or wronging to say sorry),
Is emotionally raw, but at least not scrambled:
She craves only the right to belong and be noticed,
To understand herself as a young person in the world,
In hope to find herself, safe in her integrity.
But when he, whom you trust to show you your integrity,
Bearing fruit by his polite regard for your dignity,
Robs you of that dignity, screws your integrity,
How will you find yourself, how will you acquit yourself
With what has been slighted, sullied and tossed aside?

The abuse that presses on the doors of reticence
Is a whispered world, where the language of selfhood
Has planted its syntax on loose-leaf sheets
Trampled in the dirt and blown around,
Round and round, round and round.

IV

A Gross of Egg Shells

The Sequel

(Twelve scenarios of adolescent development, based on the lack of love, understanding and stability experienced in childhood.)

From this awkward age into theirs.
Sadness splintered twelve ways
Goes forward to metamorphosis and swan song.
Impatient bones bursting through skin
Fly into the arms of the companionable day
And relapse with hardly a pang to the loved
Objects of childhood. Souls unknown
To any fostering Muse mutely hail
The incisiveness of deeds, of natural mastery
And proofs of physical and sexual prowess.
Hearts slowed to a gradgrind conformity
Now burst blood-vessels to avoid catalepsy
And to find a self in the big, wide world.
If young'uns don't sublimate self-proving attitudes
Through sport, vocational or martial training,
Hiding their humaner image behind uniform,
They may toughen by being able to bond empathically,
Capitalizing on parents' soft-hearted investment.
Theirs would be the body of trusting relationships
With parents, peers and extraparental grown-ups,
From which they draw the sinews of their strength
To get through and to suss life. Carrying wealth in themselves,
They may just resist the cynic's gritty realism . . .

1. 'Later, dear . . .'

'Dodo' had no choice in the pills prescribed
For her nerves, they were Dad's – but not her new name,
'Withdrawn'. Aloofly sulking with Triceratops,
She made her terms with rejection, and defensiveness
Became her shield. Her father provided her
With a soberly unnurturing model to emulate,
Which her girlfriends magnified as 'playing hard to get' –
That calm unresponsiveness to signals of affection
By which she re-enacted her inevitable rebuff.

2. 'Straighter, dear . . .'

The child of overbearing parents grew up
Into a mockery of the person he was, a crippled
Scholar. Indecisive, he has known no truce
With thought, murderer of the witless impulse.
But enigmatic thought pupated from timidity
And diffidence to egomaniac trust in his feelings
And dared imagination to lead the way.

3. 'Greater, dear . . .'

It was her parents' perfectionism, not swotting,
That soaked the promising academic air
With the sense that she could not win. Success
Offered no solace, failure no light
On ways of improvement. Where had she mislaid
Her competence that she never was praised, never
Could do well enough, but when she 'failed'
She was made to feel helpless? Demoralized and confused,
She grew to harbour a dislike of studying
Which had earned so little intelligent approval,
And defiantly sought from her peers age-appropriate
Life-skills, not found in her schoolbag. To one
Who was not sporting or deft-fingered, friendship was academic
Or green-eyed, and no hero figures could satisfy –

So she hardly knew who she was. But 'they' knew.
They would decide what was fitting for an adolescent –
They, who kept their brains somewhere else.

4. 'Beta, dear . . .'

The girl who grew in her brother's shadow
Saved her self-worth's last sad remnant
By doing what she knew he could never do:
At 14, she got pregnant – then aborted her remnant.

5. 'Deflate-a-dear'

How frail a devalued child is! But adolescence
Has a surprise in store for parental no-hopers:
Impotence of soul may gain a perverse sense
Of mastery from meriting disparaging labels!
Could the boy, actually bright, forgive the stupidity
That they were bent on committing in him, and lose
The chance of vengeance in his soul's suicide?

6. 'Castigate-a-dear'

The pitiable asks no reassurance that cruelty
Is merited: he feels, not weighs, the crime.
The child's own sense of his humanity is unmade
Over and over, but the liberty of self-pity
Rebuilds it. But soon he regrets the deformity
Of tears and seeks the springs of compunction
From 'taking like a man' the stripping of his humanity:
He seeks to be at one with violent abuse
By dehumanizing himself – 'I deserved that thrashing' –
As if he were a thing. *De facto* reductionism.
So grows the small and naturally sensitive
Into an inhuman object that admires its tormentor.
Predictably, the egg that was broken crushes
Itself, then visits its forsworn humanity
On qualified others, like the bullied and Pakis,

Whom it easily dehumanizes – and then its kids.
At first, they deal with their tension by crying.
Would their tears contain or attentuate Dad's anger?
Did his own? How casually, then, to exorcise
The demon fear of rejection, do they seek
To bury their faces in his godlike retribution,
Justifying the punishment – as slowly the poison
That strips their humanity fills their bloodstream.

7. 'Property allocator, dear . . .'

Not so remote from Tarquin
Were the ragged veils that sundered playmates
In the high-rise flats
That echoed his prosperous seclusion –
All the way from the wrong side of town.
Bare feet crossed the floor of his childhood
From an exotic imager that he could tune into,
And forbidden knowledge became his consciousness
And reason for coping. So he learned to condescend
To his patrons, did regal Tarquinius Superbus,
Expelled for despotism, a tribute to his good sense.

Samantha read in hard-won smiles
Sneers in keeping with her insufferable demands.
She viewed her heart as a hostile land
Full of longings and joys not admitted to;
For they gave to her loving and importunate lips
The increasing power to earn a rebuttal
That would seal her snarl and destitute fate.
So Futility struggled in phantasy to hide
The roar of her lovingness, the growl of her desires,
To disown the liberty of expressing herself,
Till finally she said, 'It's not 'I' who am excited,
'I' who am angry, disappointed and frustrated,
But another needy self, by the name of Schizo.'

8. 'Donator, dear . . .'

There are three kinds of frugal outsider:
Mum on her travels, briefly needing,
Perennially desired; the adaptable Travellers
With their familial culture;
And the unemployed housedweller.
Kids on the move bond to no screen
But to kin and neighbours sprung from the earth –
A belonging network as meaningful, as consistent,
As steady as any field; through the Family School
The resourceful pass, and emerge as routefinders,
Economists of local supply and demand,
Ingenious opportunists in occupational choice,
Psychologists of people's needs and weaknesses,
Face-to-face salesmen with the gift of the gab,
Scrap-dealers, tarmackers, hawkers, pickers,
Turning their hand to anything. Questing
Literacy, to silence the mouths of meddlers,
They sell back our leftovers and tell our fortune:
Against Travellers' Craft the possibility of housedwellers'
Long-term insecurity in a kin-free zone.
The tottering State, the loitering aimless
Evince the lack of a family-based economy.
For what will the son of the unwaged be doing
Far from Nature, with his settled ideology?
Greeting his mother the earth, his brother
The sky? Preserving his cultural identity?
Will he not lose face? – lose heart, in turn
Becoming a statistic in the unemployment rate?
How will he feel when the grim reaper calls
And life and productivity are smeared in the dust?
Will our society's ethos and communicable soul
Teach him to value his creative potential?
Or will greed be internalized as acceptable philosophy
And hollow consumption as enviable life-style?
For a golden razor is dehumanizing the poor,
Shaving the heads of alienated felons,
Notching up a street trade in Lethe's waters,

Cutting the sober with 'Back to Basics' ideas,
Slicing the heart right out of the family.

9. 'Deus Pater, dear . . .'

He made their families like flocks, without honour
In their pens if they hadn't the heart to gather,
Or as a field of wheat, gently waving
In the house of one, in the house of all;
And Paul said, 'Parents, never drive your lambs
To resentment, but correct and guide them as He would.'
So the hand that guides is the hand that cherishes,
And the hand that cherishes is the hand that leads.
But the bruised reed in due course snapped,
The smouldering wick went out at last,
And one of His little ones, the despised little ones,
Strayed – and wondered upon the Mount of Olives.

10. 'Waiter, dear?'

By the contentment on the face of the well-fed infant,
By the death of the empathic and the birth of the nonchalant,
By the one night stand and the eclipse of the gallant,
We salute the advent
Of the stuffed adolescent.

By pandering to the impulses that make baby smile,
By always being there lest our absence rile,
By peace at any price to subdue the volatile,
We hasten the senile
In the bored juvenile.

By the baby on pot, amiable at thirty,
By our feet of clay that walk the cherty,
By consumptively courting the salesmanly flirty,
We inure the shirty
To our extended puberty.

11. 'Desiderata, dear . . .'

His father's influence had been a daylight song
Of vocational and sexual success.
His father rose out of sleep to start each day
(As may a teacher, preacher or youth leader)
As the unfolding tale of the boy's masculinity,
To add wings to his dream of being someone like him
Honoured and honourable –
To provide lusty wings to carry him
High above the crudest meanings.
Between the upward-straining pinions of the self,
Fledged by models within and without the nest,
And the counterpane, rising and lunging
Like a furtive question-mark over a man's defining,
Integrity is fixed.
However reductionists may writhe out
The endless excuse of biological need,
A firm sense of self, hard-on integrity,
A respect for the dignity and integrity of others
(Which is the first part-root of our humanity)
Are ideas unknown to Biology;
They are holistic,
They don't shrink the brain
But flood it.

Tears of laughter
From TV sitcoms
Are always running down
The imprisoning petaloid walls,
That is, the softly gleaming arms,
Of the sempiternal lotus –
Down into the cynical heart of the marriage-bed.
A suddenly rising wind on the lily pond
Sent questing ripples out
On the sleep of childhood
In search of the meaning of masculinity.
They echo-located no respected males.

12. 'Traumata, dear . . .'

Darkness germinant with the power of denial
Is routed by a dawn so bright it stabs eyes
To become the dagger's instrument against mother
Who, divorcing, wasted or through fingers let slip
The boy's guarantee of a positive male influence.
But sunrising eyes may become the soul's organ
To pine untimely for the departed father,
To weigh down development with a heart of stone.
When the dispossessed teenager looks at the holes
Torn by hypocrisy in the socializing net –
In the tenuous patterns of unlovely belonging
That he'd never entreat for his sense of self –
The hostile will study Aggro-culture with his peers,
The derelict Pharmacy with benumbing zeal.

A daughter, traumatized by an absconding father,
Still keeps the parent with whom she identifies,
But rethinks the importance of a mother's moulding
When the mothering niche she scoops out for herself
Does not fill with the crumbs of a father's approval.
The girl who balanced her maternal feelings
With his smile (till it vanished) must feel her role-play
Despised and denigrated. Fierce desires may well up
To avenge maternity by disowning the rat,
Returning to sender his letters in shreds.
More power to her elbow! Love-objects, beware:
Don't be surprised if, walking out
On your duty of love, you turn into your opposite!
Daughter, far better thus to avenge
The champion of tenderness and harmony in the home
Than to accept her depreciation, grow cold and unyielding!
It is permitted for you to emulate your mother.
It is different for a boy, who must leave her to wive,
His father's name written on his forehead –
The name of one who walks out on women.

Adult estrangement is one violation of selfhood,
Extreme proximity another. Something
In a daughter's quiet is inwardly corrosive,
Something tinged with sordidness and cruelty.
She had drowned her memories in a sea of denial –
To trouble her from time to powerless time
With loathsome images floating on her conscience,
As fresh at eighteen as if it were yesterday –
But yesterday could be handled. Why is she sad?
That is the unanswered question, where relief
Is only found in unburdening the joyless
Child in the adult no one will listen to.

V

The Elm and the Vine

(One-third of doomed marriages dissolve in alcohol, the rest in a weak solution of infatuation, expediency, desire for a family, for status or economic security, or are attacked by the acid, infidelity. The family is the developmental setting for a child's education in a love that aims at integrity – i. e. the integration of opposites to subserve the couple's erotic needs, the child's nurturing and society's recruitment. Recognition of the importance of balancing these conflicting interests leads to an appreciation of three requisites for a child's development into an integrated person.)

Picture colts and fillies freely grazing
In a field where stands a dignified English Elm:
In outline like the billows of a thundercloud,
Meagrely seeding (it spreads by root-suckers),
Bearing dud fruits, erratically, moodily,
Its few statuesque ascending limbs
Twisting in upon themselves
And around themselves,
Fizzling out fast in greeny-black dense domes
Constructed on slinky, twiggy curlicues.
Now the incurved and self-centred elm falls in love –
With a grape-vine that grows on a wall,
Climbing with her bounty to obscure all property.
Neither plant lacks the confidence to be itself:
The lovers keep faith with their own identities
The better to give of themselves to the other.
The elm learns the viticulture of his soul,
The vine, in time, the silviculture of hers.
The elm seems to defy gravity, massively soaring
From its heart, the straight bole persisting
Right through the crown, simple, open, positive –
And so his love is, and spreads itself around,
Bearing the weight of all sprouted cares.
The elm is the strength and glory of the vine,

The vine the mellow fruitfulness of the elm.
So liberating are their reciprocal roles
That 'Strength through joy!' is soon his cry,
'Fruitfulness through patience!' becomes her chime,
As love seeds their natures to bring forth
The dream of their own true selves
And the dream of colts and fillies in a field.

1. Smother

From the fullness that speaks of love
The scene changes to half-tone human tragedy.
A vinous woman clings, but is not held –
And, in hope to be held, turns sighing to her bottle.
Another buries her soul like a memory in the man-child
She married; finding no wifely destiny,
She satisfies her mother-love in the child-man,
Protracting it well into his Oedipal years,
Blinding his eyes to the rival sex,
Superfluous to him, whom she clasps
And surrounds with futurity and possibility
Only so long as he completes her life.
Marital bliss might have saved the elm
From the fungus that rots ambivalence to daughters:
He crossed the boundaries of her ego's workshop
When he declared his reliance on her heart's resource.
She also, suddenly nurturant, had crossed the divide
That fenced her from rivalry with her own mother
Who now eyed her jealously and swore she gloated –
Though what burned and raged in her daughter
Now absorbed the energies she needed for growth.
As she grew in father-love, she turned her back on
The concomitants of horizon-extending emotional engagement:
On her problems, questions, dreams, cares, regrets,
On boyfriends, loves and losses, her self-esteem,
New relationships, new experiences, exotica,
Fresh air, deep breathing,
Life.
Years later, it was the unwept dead

Who cared for the widower.
The generation boundary permits free grazing
Where selves may grow to fuller selfhood:
The man who crossed it was no stable lad himself
And yet condemned his filly to a stabled future.
And the mollycoddle? What was his fate?
His prolonged dependency nursed the schizophrene.

2. Pother

Nurturing, educating and exemplary
Grows nurtured, possessive and insecure generation
With the sunset on wedded love.
Sunrise on uninvaded growing-space
Finds foals self-defined, self-defining.
But two suns – and shadows dissipate in confusion!
If there's discord in the heavens over rules and standards,
Or one sun lights up the other side of the world,
The child may suppose his problems are atomic
And, fearing the fall-out from conflicting directives,
Will adopt the partisan's guile to obey
Or the scapegoat's role to be blamed,
Till, all energy spent,
He will sicken and die from divided loyalties.
A preferred parent is a boon to young fantasists:
His fairy godmother is not better known
Than when seen beside her amphibian creation
In the delusive pool of a disintegrating coalition,
Where keeping the heart anchored in one parent
Does not help the son to ripen from ambivalence,
Finding no repose but perfect incongruity
In that variability-in-unity, Nature's way.
Mutual support and mutual confidence
Dub family leaders as persons of worth,
By a common life united, not blended or submerged;
And when parents play a duet, the children chime in,
Seeking mutual allowances and mutual understandings.
This is the family song,
The song of integrity,

Where each has learned to accept the other
For what he is and for what she is
And for what each may become,
Given love, understanding and patience.
The sweetest airs were ever harmonies.

3. Other

The family was not otherwise designed than for peace
And consensus. Then children may enter its heart
And make it a house, filled with light.
Outside, a world of different circumstances
Prescribes their roles (or else their penalties),
The garb most fitting for their time and place,
Their abilities and characters, their likes and dislikes –
But, like off-the-peg suits, they don't fit well.
A good tailor can alter *other*ness to fit thisness,
Adapting their costumes without compromising style –
And some, by shedding pounds, could do themselves proud –
But a bad tailor fits children for roles in life
Which so plainly are not his or her style
(Brutalizing the sensitive, putting artists behind a counter)
That they might as well be in fancy dress. –

And no roles confuse quite like gender roles.
Clothiers should know which is his or hers –
And comely are clothes appropriate to gender –
But who said, 'That battle-dress suits you, ma'am'?
Did an unromantic father or a cold mother?
And who schooled the boy in his mincing gait,
Haunted by questions no weak Dad could answer
Or dominating mother, who sexed down her son
By demonizing Romeo and banishing Juliet?
Sexist I'm not – I'm for flattering gender-difference –
But, while some are undoubtedly compromised by nature,
Others have compromise thrust upon them.
If the soul-garden blooms for the effeminate and tomboy,
They have found their true selves. But what is certain
Is that the young are suggestible and easily impressed –

And, arguably, the wife-beating football idol
(As offside in his sexuality as those who are camp)
Is no example to boys of what manliness is.
Wholesome actions should speak through roles.
Gender roles should hold us together,
Give meaning and roundness to our lives,
Free our perfection,
Be the beauty of our psyche's balancing –
Or they are not the rocks we think we build on,
But the cues we take to resolve our confusion.
If a child does not know what being
A boy or girl should mean,
If one sex does not respect the other,
But the other is all the same (debased) thing
And we are only hungry for a bit of the other –
Boys will never become gentlemen,
Girls will never become ladies,
But those parts are all played out
And only pantomime remains.

VI

Mentor

(For Dr Pierre Watter)

(Preamble: The first four books in Homer's The Odyssey tell the common folk tale of a son in search of his long-lost father. Before Odysseus, King of Ithaca, left for the ten-year Trojan War, he entrusted his affairs and his new-born son, Telemachus, to a family-friend, Mentor. After the fall of Troy, Odysseus and his crew, having been blown into the uncharted myth-world, where they blinded the Cyclops Polyphemus, son of Poseidon, are hounded for another ten years by the avenging sea-god. For Telemachus events passed smoothly till his seventeenth year, when 108 suitors invaded the palace to woo his mother, Penelope (a model of constancy, discretion and high-mindedness), in the hope of depriving the immature son of his succession and consuming his inheritance (which he was powerless to prevent). Only six weeks before Odysseus in fact returned, Athene (goddess of the civilizing arts) assumed Mentor's guise to accompany Telemachus on the first leg of a journey she instigated to seek news of his father – from two Trojan War veterans, King Nestor of Pylos and King Menelaus of Sparta. Nestor had no news of Odysseus, unlike Menelaus who, having been stranded on Pharos in Egypt on returning from Troy, had to force Proteus (a seal-herding sea-god, capable of taking on many forms) to prophesy a safe route home for him. This protean seer, who had the measure of life's ruthless changeableness, had caught sight of Odysseus alive but a prisoner on Calypso's isle. Life was sufficiently haphazard to allow Telemachus to hope that his father, like Menelaus, would soon find his way home. Thanks to Athene's advocacy, he was almost at the palace portals. Avoiding an ambush laid for him by the suitors and offering refuge to the seer, Theoclymenus, from the oppressive avenger of blood, Telemachus has now assumed the high profile, the mantle and the responsibilities of true, heroic manhood. He returns to Ithaca, where the Warrior-maid, steel-eyed Athene, again in Mentor's shape, helps the reunited father and son in their stand against the suitors and in making peace with their relatives.

The Telemachia may be viewed as a coming-of-age tale, and Mentor as the archetypal friend and sage adviser. It is not this poet's view that heroism should be associated with bloodshed or that the appallingly bloody progress of civilization should be played down one iota, but that true heroism is within the reach of all, is a legitimate aspiration for all – while mentors are very hard to find.)

'My mother says I am Odysseus's son,
But, of course, I wouldn't know. For no one
Really knows what his father stands for.' (1:215)

So sang Homer, that bard of renown,
Of Telemachus's quest to find his own identity
In awareness of his parentage, his heroism to crown.
Athene respected her suppliant and took pity
On him whose moping was the warder of reality
And whose futility and fears grew monstrous as a crag,
And said, 'All weakness is potential vitality
In need of discovery and a meaningful tag.

'Shall suitors in the midst of the stately scene
Rouse up storm-winds against you, barring
Your path on the sea of many counsels keen,
As on the cheerless strand, humming and ha-ing,
You grow fathomless deep, inert and passive,
Charted with the shoals of the gods' bitter spite?
Or, harboured in friendship, will you begin the narrative
Of your voyage of discovery and set sail tonight?'

So Athene made the heart in him strong to bear;
And the glance of his eyes called to mind the blest,
The wit was Odysseus's and the temper was there
To turn impotent whimsy into the wistful quest. –
Yet the winged words sped from lips that had bored,
From his teacher's of old, discipline's horseman,
Yet wise-witted by the grace of the Aegis-lord,
Far-seeing Zeus, like dragon-slaying Jason!

Mentor and Telemachus communed of their sorrow
That man and life's gifts weren't exceeding fair
But mean, neglected or imperilled by the morrow,
By the book half open, the cage and the snare.
But a few poor honours remain to mortals:
To laugh at the ladder and Charybdis's whirlpool,
To cleave the murk, safe among pals,
To be serene in the doldrums and of Zephyrus worshipful.

So philosophic a frame as heart-to-hearts beget
Was theirs as they sat and just tarried awhile
In the house of true-hearted Penelope, to let
Minds drift like the ship of the wise-souled exile
On the foaming sea, so deep. How long
Would Telemachus stand aloof from being that hero
He had not yet dared to be? How long
With him, also, would dally the Concealer, Calypso,

She of the fair locks, shackling the coy?
Very soon he would break the soft bands of sleep
And be wafted from the Nymph, that enervating killjoy
Who wouldn't allow him to sail his own deep,
To find out for himself what a son can do.
Keen winds of healing amply breathed
From the challenge he took up, the opportunity quite new
To know a world of heroes and his own place, bequeathed.

VII

The Faithful Swineherd

'Here I sit, yearning and mourning for the best of masters and fattening his hogs for others to eat, while he himself, starving as like as not, is lost in foreign lands and tramping through strange towns – if indeed he is alive and can see the light of day. But follow me, old man, let's go to my hut. When you have had all the bread and wine you want, you shall tell me where you come from and what your troubles are.'

(The swineherd, Eumaeus, to Odysseus, in disguise.
From Homer's The Odyssey, tr. D. C. H. Rieu, ch. 14, ll. 40–7; Penguin, 1991)

May Zeus forbid such a thing should befall
So contrary and brutal as befell King Ctesius,
Whose son was kidnapped by his nurse when small
And bought as a slave by the father of Odysseus!
Woe to be snatched from that hearth where he thought
He swayed the sceptre over a world of good,
To be gathering the buds of content from aught
That his master Odysseus felt that he should!

This life of manifold shifts undefines us –
And for this the swineherd bore his parentage in vain –
But in the house of our thrall, if dejection finds us,
We fare less well than the beloved swain
Who o'ermastered his servitude with the courage to be.
Mourning, but lulled in the arms of Odysseus,
Eumaeus redefined himself, not as compassed with misery,
The victim of circumstance, but heir to a spontaneous

And unquestioned tradition, where no one winks
At the keeping of slaves or misprizes tokens
That the master is friend. The ghost of a jinx
Was laid with the childhood illusion of omnipotence;
And his 'I' was not lost in that process of consoling
Or in the fullness he found in commitment to duty,
For decency prevailed over self-interests's cajoling
And the conventional self over mercenary incivility.

Lest Freedom's daughters at Homer should rail
If wronged Eumaeus too lightly consented
To custom's lordly will, I shall detail
The incidental things that prove him contented:
The plan for farm buildings, deep in his mind,
The courtyard wall, a labour of love –
To contrast with our duties, minutely defined,
Not risking intimacy and the spontaneity thereof.

The despair-witted workers, who never uncoil,
Are freemen today! Our advantage is gained
By trying *not* to lose ourselves in our toil
And fool our hearts that no choices remained:
A too eager role-play is the doom of self-consciousness
And the vengeance of self-interest is hard by the door
To accuse the suppliants of work's meaningfulness
Of surrendering their freedom to work for more!

High flyer, upborne on the wings of plaudits,
Toil-addicted, claiming the right to success,
Icarus plummets from the heat of his audits:
The misuse of choice in the quest for stress!
Can the meaning of life to his mill be grist
That he uses his vaunted freedom to smother?
Does he know who he is, as the swineherd wist
In virtue of his odyssey and his relation to another?

VIII

Persona non grata

'"Sir," said the goddess of the gleaming eyes, "you must be a simpleton or have travelled very far from your home to ask me what this country is. It has a name by no means inglorious . . . the name of Ithaca . . ."

Odysseus' patient heart leapt up as Pallas Athene, Daughter of aegis-bearing Zeus, told him its name, and he revelled in the knowledge that he was on his native soil. He addressed her with words on wings, but not with the truth. True as ever to his own interests, he held back the words that were on his lips.'

(from The Odyssey, tr. D. C. H. Rieu, ch. 13, ll. 235–9, 250–5; Penguin, 1991)

O herald (but no angel) of the gods, Mercurius,
What a prize for your prowess as the wiliest of thieves
Was your cunning confidant, Autolycus, Odysseus's
Grandfather, who gave him the wit that deceives
And his name, the Antagonizer! Could the mind of Odysseus,
As it fed on subterfuge, sail into consciousness
Of his own true heart, known to Eumaeus,
After twenty years' learning the ways of toughness?

Slippery as Proteus and as hard to pin down,
His identity, when a guest, shifted in fun
From outcast to sucker, from coward to clown,
From no one who became someone to someone who became No one;
The indulgent captain could switch to the mischief-maker
Whose relentless curiosity laid his men low;
He was the sociable stickler, the personable faker,
The reticent fun-seeker, the pious ego.

But later among the suitors in a beggar's guise
He prayed: 'All-father, begetter of heroes,
Today grant restraint and victory to the wise,
But tomorrow to the true heart give the lips that disclose
The truth; for I blinded the man-eating Cyclops
And no bestializing Circe remains to seduce,
To exchange the meat of self-awareness for slops –
And to Penelope I came home, high-minded as Zeus.'

Once his piety was stoical in its fraudulence:
He spoke no hard words about conscience guiding
The seeker over measureless tracks of consequence,
But of accepting one's lot under nature's presiding,
Prosperity or adversity from the gods in good heart:
'Let no man be lawless or squander his gifts,
For life is uncertain and wrongdoers will smart
From the whips of Providence and life's many shifts.'

Odysseus in his wanderings had learnt life's irony,
The helplessness of man, the awesomeness of nature.
Hence few the threads of his counsel span he,
Compared with the ropy speculations on the future
Of the halting who weigh acts by entailments. In a veiling
Mist he shrewdly simplified the world
It were more for his glory to be with arrows assailing
Than questions about futures into darkness hurled.

In a bleak soft focus he got his own way.
But his wary disabuse still o'ermantled Ithaca,
A strange land on returning, till clear-eyed Athene
Purged the demons from the consummate tricker
And dispersed the mist. He knew himself then.
He stowed his cynic in a cave to spin gold
For two thousand seven hundred years. Amen
To homecoming. 'Amen!' echoed all his household.

IX

The World Your Heart Partakes Of

The world your heart partakes of,
To meet the needs it aches of,
Science quakes of.

The realm on which consciousness feeds
Borrows some hue from needs
That no scientist reads.

Abstract, ageless reality,
Scientists' revered majesty,
Is no accessory

To your private world of experience,
Ever-changing and investing with ambience
Your frames of reference.

The whole Earth approves when you frame
A consummatory niche with your name
(No two the same),

When your mind, not skittering with the times,
With the immensity of the cosmos chimes,
To rest betimes. –

And the sky can hold your blue,
The buzzard can cry to you,
Significant, too.

You are because you belong,
Your perceptions are singing your song,
Your reality's throng.

X

Multiple Choice

Who is the happy carer?
She, who her own life carefully scanning
Thanks heaven it fared far better for her? –
Or, who gladly sacrifices that valueless thing
To allay the canker at her heart,
Her fear of the dumpcart?

Who is the willing horse?
He, who not daring to ask for the sky
Between work and play runs a positive course? –
Or, who sets his standards impossibly high,
The fear of poor grades so earnest
In the mind of the perfectionist?

Who is baring her soul?
She, who speaks as if writing her diary,
To give value to experience on her own terms her goal? –
Or, who breathes the deprecating air of the priory
From childhood, to keep her counsel –
For her therapist's goodwill?

Who really owns himself?
He, who choosing the destiny assigned him
Upholds the values consistent with self? –
Or, self-disavowing when weakness finds him
Denies authorship of the wrong
He condemned for so long?

Who comes to know themselves?
She, who downcast likes him who loves her,
Who knows her for the unfathomed mystery he delves? –
Or another, not sharing his concerns with her,
Who makes a granite spouse
And a well-run house?